Where Theology and Science Overlap

DR. MICHAEL FITCH *and* **GEOFF HAWLEY**

WestBow Press books may be ordered through booksellers or by contacting:
WestBow Press
A Division of Thomas Nelson
1663 Liberty Drive
Bloomington, IN 47403
www.westbowpress.com
1 (866) 928-1240

ISBN: 978-1-4908-1010-2 (sc)
ISBN: 978-1-4908-1009-6 (e)

Library of Congress Control Number: 2013917600

Printed in the United States of America.

WestBow Press rev. date: 12/9/2013

HI ELIZABETH,

IT THRILLS ME TO HEAR OF YOUR DESIRE TO SERVE THE LORD, NEVER GIVE UP! THERE IS ALWAYS MORE WITH GOD. THIS BOOK IS A RESULT FROM 9 SEMINARS WE WERE ASKED TO DO ON 'CREATION VRS EVOLUTION'. (DURING THE FIRST SEMINAR A GUY CALLED JOHN GAVE HIS LIFE TO JESUS) I PRAY. THIS BOOK WILL ANSWER A FEW QUESTIONS BUT ALSO CAUSE YOU TO ASK MORE.

LOVE IN CHRIST,

[illegible] 24-4-14

GAL 4:4-7

Take courage – you can believe the Bible!

God bless, Michael [illegible].

Ex 20:11

Dedication

This book is dedicated to three groups of people.

First, to those who are interested in exploring the reality that is beyond the physical senses.

Second, to our friends and fellow Christians who we hope will be strengthened and have increased confidence in the Biblical creation.

Third, to our families, especially Matthew, Rachel, Emma, Joanne, Jonathan and Saphiere. We hope that this book will encourage them in their personal search for fulfillment and understanding.

About the Authors

Michael Fitch was born in London in 1955. He was educated at Kilburn Grammar School and over the course of quite a few years gained a diploma in electronics, and a first degree and PhD in physics and mathematics. His working career is research and development, and he has worked for a number of communications companies. Michael became a Christian in 1982 while living temporarily in New Zealand and working at a company that had a number of Christians on the staff. He debated with them for more than a year, dismissing Christianity as a curiosity and very unscientific. Then one night he had an encounter with God that changed his life forever. Since then he has become increasingly interested in the way the universe works spiritually as well as physically. He is of the opinion that there is no conflict between science and the Christian faith. Indeed he has become increasingly convinced that all scientific observations support the Word of God as recorded in the Bible. Over the past five years, Michael has given talks on evidence for creation and design at churches and anywhere else that would hear him. Michael is married to Dawn and has a son and daughter, and is a deacon at Ipswich International Church in Suffolk, England.

Geoff Hawley was born in Ipswich in 1956. He attended Westbourne High School and went into a career in sales. He was brought up in church from the age of two and became a Christian at the age of seven. Even as a boy he somehow knew that all the answers were in the Bible. Geoff's mum spiritually supported him and his Dad nurtured an interest in biblical prophecy. At the age of sixteen he compromised his faith and then, following a near-death experience where he spent twenty minutes going down towards Hell, Geoff spent a further two years struggling with commitment and guilt. Finally God caught up with him after he listened to a recorded teaching on the second coming of Christ. He became a deacon and then an elder at Ipswich International Church. Geoff's special interests are the Word of God, covenants and prophecy. He is married to (another) Dawn and they have two daughters, a son, and a granddaughter.

Foreword

Faith in the Christian Bible and particularly in the Genesis account of creation and Bible prophecy in this scientific age demands moral courage. Yet we have Jesus the author and finisher of our faith as our great forerunner. He did not only believe the historicity of the creation story; he vouched his life upon the infallibility of the scriptures and fulfilment of biblical prophecy as recorded in John 10:35 and Matthew 5:17-19. We make nonsense of all Jesus's convictions, His sermons, His death on the cross and our salvation in him if we throw doubt on the verbal plenary inspiration of the scriptures according to Matthew 26:24.

To attempt the marrying of macro-evolution theory with the Bible account of creation is foolishness at the best and self-deception at the worst. Like water and oil, the two can never mix. Micro-evolution is no new scientific discovery. Absalom rode a mule to battle in 2 Samuel 18:9.

Bible prophecy is another area of disagreement. Some churches do not mention it; others see scripture fulfilled in every news item and tell us the day and hour of Christ's second coming! The fact that over 300 Old Testament prophecies were literally fulfilled in the first coming of our Lord Jesus makes it dangerous to ignore Bible prophecy concerning His second coming. We may not agree on all counts of interpretation of these yet-to-be-fulfilled prophecies, but we have reason to respect His words. Jesus said He would come born of a virgin; He came. He said He would die crucified; He was hung on a tree. He said He would rise the third day; He rose. He said He would go back to His Father in Heaven; He is now seated at God's right hand. He said He would come again; I choose to believe him!

Covenants, Creation and Choice is a timely book for Christians struggling to hold the Bible and science together. Here, Dr. Fitch and Mr. Hawley argue that it takes more blind faith to believe in evolution as the source of life on Earth than the Genesis account of a six day creation by a master designer the Bible describes as God the Creator. They persuade us that the Covenants of God have life-transforming consequences and that Bible prophecy merits serious attention.

For all those shattered by the apparent contradiction between scientific speculations and the Christian faith, this book is welcome news from the desk of a qualified scientist who dares to interpret the facts differently. There is meat here to chew for all who wish to make sense of Biblical prophecy and the covenants and their relevance to the times we live in. Each chapter feels like another cave in a treasure hunt.

JFK Mensah
Chairman, Great Commission Church International.

Acknowledgements

The authors would like to thank the many people that have helped and encouraged them during the creation of this book. In particular, we would like to thank our wives and families for being supportive, and tolerating us spending many hours studying and writing. Michael Fitch's wife Dawn is an artist and, together with Geoff Hawley, provided many of the illustrations. We also thank members of our church, especially the senior pastor Reverend Harold Afflu, who reviewed much of the text and made suggestions for improvements. Thanks are also offered to Reverend JFK Mensah who has also reviewed the manuscript and made many suggestions for improvements, especially related to Christian doctrine. He also kindly provided a foreword. Responsibility for the text and opinions expressed is solely that of the authors.

Contents

Glossary

Contract	A contract is an agreement between two parties that is negotiated. It differs from a covenant, in that a covenant is non-negotiable.
Covenant	A covenant is an agreement between two or more parties. Biblical covenants come in two varieties, unilateral and bilateral. With a unilateral covenant, God imposes the covenant and terms and conditions associated with it. With a bilateral covenant there is commitment from both sides. A timescale is given for when the covenant is in force. A bilateral covenant has one additional ingredient when compared to a unilateral covenant: ratification that involves the sprinkling or shedding of blood. A token or a sign is often given by God, and a celebration meal is usually held.
Creation	The means by which the heavens and the Earth came into being.

DNA	Deoxyribonucleic Acid is present in all living cells. It contains the genetic information for the construction of all the different cells in a body.
Evolution	The process of living organisms improving their probability of survival by adapting to changes in their environment. There are two kinds of evolution mentioned in this book: micro-evolution and macro-evolution. Micro-evolution is where a species adapts to environmental changes by losing some functions or the environment provides natural selection of characteristics like colour or size. Macro-evolution is where a new species is formed from another one (or other ones).
heavens	There are three layers of heavens referred to in the scriptures; first and lowest is the sky, the second is space and all it contains, and the third is the highest Heaven where God's throne is. The lower two layers are physical in the sense that they, along with the Earth itself and everything in it, comprise the universe. The third Heaven is spiritual (or unseen), outside the dimensions of the universe.
Heaven	The highest Heaven, where God dwells. Also the present location of Paradise.
Mankind	The human species, with no differentiation between men and women.
Number	Numbers are important in the Bible and they have meanings. Certain numbers mean certain things. One, two, three, five, seven, eight, ten, twelve, forty four hundred, and one thousand, are the most commonly

used. Number one means God, number two means agreements (as in covenants) and number three represents the Trinity. The number five is usually associated with grace. The number seven is sometimes ascribed to completion (sometimes final) as meaning perfection. Multiples of seven are found throughout the Bible from the creation account in Genesis through to bowls of wrath in Revelation. The number eight is Jesus's number; it encompasses the overflow, blessing, and new start. Number ten means wisdom, decisions, and responsibilities. Number twelve represents government and authority structure. Numbers forty to four hundred are associated with time of testing or judgement, discipline, and sometimes patience. One thousand is often used to denote power as in property, but is often used to denote a very large number when the number itself is unknown or is infinity. In Bible language words, male and female and letters each have a numeral value, and studies have been made of the significance of numbers represented by Hebrew writing.

Sin — Sin has three classifications, which are trespassing, iniquity, and transgression. Trespassing is encroaching on someone else's space without authority, such as when someone enters private property. Iniquity includes the concepts of scheming and planning as well as the sinful outcome from it. Transgression is a term used for the actual breaking of a law.

Symbol — A symbol is a representation, a proxy, standing in for the actual information. For example, a number may represent information as a pre-arranged code.

Token	A sign or a symbol that serves as a reminder of a covenant.
Trinity	The three persons of God, who are the Father, Son, and Holy Spirit.
Type	A type is a precedence that sets a pattern, or that follows a pattern already set. For example Melchizedek is one of the types (in this case of a priestly order) that Jesus later followed.
Universe	All of the material creation, both seen and unseen. This comprises the Earth and everything upon it and within it, and the lower two layers of the heavens, which are the sky and space.

Introduction

It is an undeniable fact that you and everyone else are here, living upon the Earth, so one certain truth is that mankind began to exist on Earth somehow. Opinions vary about how we got here depending on what you believe. The range of these opinions is very wide. At one end of the range is the belief that we came about as the final part of a creation process that took just a few days about six thousand years ago. At the other end of the range is the belief that we came about through evolution over millions of years.

We are positioned at the first end of this range. We believe that the Earth was created about six thousand years ago, over six literal days, as described in the Bible in the book of Genesis. Combined with this belief is our certainty that God works through covenants and has an overall covenant with His creation. This covenant has been through several stages of modification and development already, and has several more to come.

We confess that we are Christians and believe that the Bible is the written Word of God, and we draw heavily upon it to uncover

God's motives and plans. The subject of covenants and their connection to science and creation has been largely ignored in other texts because we believe it is not well understood. We found that once we began to research and understand God's covenants and combine this understanding with science, the evidence for creation and a young Earth became even more compelling.

The aim of this book is to present arguments and evidence for covenants and creation and to explain the choices we have before us that will determine our destiny. The realisation that God is the Creator and has a covenant with His creation is truly life-changing, as if suddenly everything falls into place. Hopefully after reading this book you will be able to make a more informed choice about your own destiny.

Like most other people we were taught at school that evolution took place over millions of years and we believed it. Many scientists sincerely believe this and will insist that creation over a few days is not possible. Some Christians also sincerely believe in evolution in one form or another.

Maybe you think the creation story is at odds with scientific evidence that has been peer-reviewed and has accumulated over decades. Of course, just because there is a lot of evolutionary scientific material, it does not mean that evolution is correct. Science can only explain the here and now, and any extrapolation back in time beyond a couple of thousand years is based on unprovable assumptions. A scientific theory can be proposed by anyone, and is scientific only if it can be tested through observations. As each new observation is made, the theory survives provided it can explain the new observation. However if just one observation is made that cannot be explained by the theory, then the theory must be modified to take the new observation into account or, if this is not possible, the theory must be abandoned. Unfortunately, the theory of evolution cannot explain all observations and ordinarily it would not be allowed to stand as science. However it has survived by being changed into a scientific 'law', which apparently makes it immune to being crippled by observations that do not fit.

We have noticed that the theory of evolution is supported by governments and the media, who take every opportunity to promote it. Nature programs constantly tell us about how life and the universe have evolved over millions of years through a series of random mutations. In this book, we have brought together observations that support an alternative theory whereby the universe and life within it came about through creation by a master designer. We assert that the account given in Genesis is correct and give some supporting evidence; we cannot hope to give it all in a single book. It is our hope that through reading it, you will be inspired to recognise more evidence for yourself.

We believe that Genesis gives a true and literal account of creation because it is part of the Bible, which is God's written Word. If we choose not to believe this part, then what other parts should we not believe? We would need some criteria to decide what parts to believe and what parts not to believe, and this criteria would be generated by humans. Surely the account of creation in the Bible is foundational to Christians. If we do not have this firm foundation, then our Christian lives will collapse; we make out Jesus to be a liar and His work on the cross is for nothing. Indeed we decided to write this book in obedience to Isaiah 58:12 that tells us to *'rebuild the ancient foundations'*.

There is an adversary whose names include the Devil and Satan, who will try to cast doubt in people's minds on this issue. Part of his strategy is planting doubts about God's Word in people's minds and the creation account is something that he is particularly keen to attack. He knows that if this foundation can be damaged then all the Christian values built on top, such as the integrity of marriage, will collapse. It is necessary to understand the role and destiny of Satan within creation to understand why events are unfolding as they are, and to understand what will happen in the future.

We warn you that this path can be a challenging one to take. When you tell people you believe in creation over six days and also in a young Earth, you fly in the face of entrenched teaching and beliefs in evolution. You run the risk of being thought of as eccentric, a religious nutcase, unscientific, and you may even lose friends. A previously good friend said to me, 'I have always considered you intelligent and I am amazed that you believe this rubbish. It is all wrong, and you are deluded'. He got quite angry about it, and it has put a gulf between us. Creation is an emotive subject, suspiciously so, almost as if people are terrified of the possibility of there being a Creator, and that we were created not so long ago.

Are you up for the challenge? We would encourage you to dig for yourself and see what you think, and not to be afraid of joining increasing numbers of people, including scientists, who reject evolutionary thinking. As the book unfolds we will point out how, with the appropriate interpretation of observations, there is no conflict between science, creation, and covenants.

Covenant thinking transforms lives. It improves our understanding of how God works and why certain things have been done (and will be done) in certain ways. This book is not primarily an evangelistic one, but the salvation message through Jesus Christ shines through loud and clear anyway. We are confident that, while reading the book, you will want to form a relationship with Him if you haven't formed one already. So beware, reading this book will change your life! If you are already a Christian, we hope it encourages and inspires you.

From Genesis to Revelation, we see that God does not condescend to mankind's level but rather enables mankind (who are God's image bearers) to approach His level through His covenants.

In God's mind, the story of mankind is like a jigsaw puzzle before it is cut up. We have been given the box with the picture on it, and our task is to re-assemble the jigsaw for our own lives, taking pieces from the Bible. This book gives clues, keys and facts that will help you in this task, to stimulate a better understanding of how God and His creation and covenants work.

We present scientific evidence in support of creation in a popular science style, which will hopefully appeal to most people, rather than as a rigorous scientific study. Likewise, we also present information on God's covenants in an easy-going style rather than as a rigorous theological study. This said, we do go deep sometimes and give additional 'points to ponder' in an attempt to inspire the more academic reader.

This book contains chapters on covenants and on science, presenting them somewhat separately but with links identified, and then they are brought together in the final chapter (chapter 10) on restoration. Along the way we try to give some insight into why God created the universe. We also try to give insight into why things are happening in the world that we are witnessing at the present time, and what will happen in the future when all things are restored. Many books on the subject of scripture explore the what, the how and the when, especially related to end-times prophecies, but this book also reveals the why.

Father God is the Master of all creation because He created everything for its own purpose from His understanding of spiritual principles and material physics. To comprehend the spiritual dimensions that God lives in is something that we mortal humans find impossible because we were created to live in the physical realm. We can however believe in the existence of spiritual dimensions without directly observing them, just as some people believe in evolution over millions of years, without directly observing it. This kind of belief is called faith. Faith is described as '*the belief in something unseen*' in Hebrews 11:3.

God entered His covenant into this world and this took supernatural means, and so it will take faith on your part to believe it. It takes faith to believe in God the Creator, but then again it takes faith to believe in the theory of evolution over millions of years and this, in our view, is a much more blind and misguided faith.

To help us in our quest to see or comprehend the spiritual dimensions of God's environment, the Bible gives us sufficient insight. God's ultimate plan for us was discussed, agreed upon and then written down. This plan was basically two-fold: first to share His love with humans that could love Him back and, second, God had to deal with the sin nature and its consequences. This second area of the plan had to deal with paying the price, buying back and restoring, otherwise known as salvation and redemption. This had to include not only mankind but also the world where we live.

It is argued by some that Genesis is not literal, but a Hebrew poetic allegory of folk-lore. But we are taking it literally as God's Word, and we shall see how its implications pan out across the sixty-six books of the canon of scripture that we call the Bible. It is our hope and desire that, through our book, you will acquire keys of understanding into God's motives and plans, and subsequently Satan's motives and plans to subvert God's plans. With the outlay of scripture we can see what God intended for, by, and to mankind (Hebrew = Adam). We see that Jesus is the mainstay reason, allowing God's love to deal with a relationship with His entire creation, from the universe itself right down to the creation of mankind in His image. It doesn't stop there; He has an ultimate plan for mankind to become spiritual and to enjoy future ages with Himself. In order to understand this, we must enlarge our thinking and experience from being constrained by the physical realm or zone into the spiritual realm or zone. There will be an end to the physical life and the start of the next phase, which will be a spiritual life. God tells us that sin kills or brings death to our physical body, which is just a shell for our spirit. The life force stops or leaves, which causes the shell or body to decay and return back to the Earth. Then the spirit or soul, which is not physical despite arguments that it is a chemical process, is drawn through the doorway of which the entrance is death.

At the end of the Bible in the book of Revelation, we have many conclusions to the issues of mankind and the Earth. The book of Revelation was almost left out in 1611 when the present books of the Bible were selected. Over many years it has continued to harass, challenge and inspire not just theologians but many types of people from many different walks of life including art, music, and literature. If it had been left out, we would not understand the Bible as we do today. For example in Revelation 13:8 we read of a book in Heaven, the Book of Life, which contains the names of people who belong to Jesus. We also read that Jesus sacrificed His life to purchase salvation for these people. Revelation also says that He was slain, killed, and slaughtered before the world was created. But wasn't Jesus crucified two thousand years ago on a cross on a hill just outside Jerusalem? Yes, but to God the plan at the time of creation was a set covenant; as far as He was concerned it was as good as done. The timing of the crucifixion was crucial, as is the timing for every event within the covenant, in order to fulfil every detail and loophole to cover the price, so that it is totally paid.

It is recommended that you have a Bible to hand when reading this book as many references will be made to it, and without the relevant scriptures you will not understand the full context of what is being said. It is highly advisable to read all the referenced scriptures. Versions recommended, in order of preference, are the King James Version (KJV), the New King James Version (NKJV), the New Living Translation (NLT) and the New International Version (NIV). The references quoted in this book are from the NKJV unless otherwise indicated.

If you are unfamiliar with the Bible, we recommend that you read certain passages to form a foundation before you continue reading this book. In the Old Testament section read Genesis chapters 1, 2, and 3 and Exodus chapter 19. This will take you through the story of creation and the promise to Israel. In the New Testament section read the entire book of the gospel according to John and Revelation chapters 19, 20, and 21. This will take you through the fulfilment in God's son Jesus, who is God's covenant token of love to mankind, and then to the final conclusion of the ages of this creation. We guarantee that you will be excited when reading these scriptures!

The Trinity

God in three persons

The first verse in the Bible is Genesis 1:1 and it says: '*In the beginning, God created the heavens and the earth*'.

The Hebrew word for God in this verse is Elohim, which is a plural word. The word is also half male and half female, but we will park this thought until chapter 6 when we talk about Adam and Eve. So why is Elohim translated as God and not Gods? It is our belief that God is one entity but at the same time represents Himself in three persons, traditionally referred to as the Trinity. The word Trinity does not appear in the Bible and was first used in the late second century. The Trinity is three representations or persons or projections of God. One possible reason for God being projected by these three persons is that it enables us to understand something about Him. The Trinity consists of God almighty, Jesus Christ and the Holy Spirit, sometimes referred to as God the Father, God the Son and God the Holy Spirit. It is an artefact of language and patriarch tradition that God is represented overall as male; in truth He is sexless.

We do not attempt to answer the question 'who made God?', or try to explain how or why God was created. We don't know, indeed no-one can know. We are not equipped to deal with the answer even if we were to be told what it is. It is an absurd question, a type of question that is not logically correct to ask. Imagine a computer made by human hands. Even if it was equipped with the latest software, it would not be able to understand how humans came about. This is because our ways and thoughts are higher than the ways and thoughts of the computer, if indeed it has any conscious thoughts. A computer may have knowledge of how humans came about, but this knowledge would have been put in by humans. Our thoughts and ways are far above those of the computer, because it was our creation. In the same way God's thoughts and ways are far above ours.

> *'For as the heavens are higher than the earth, so are My ways higher than your ways, and My thoughts than your thoughts.'*
>
> **Isaiah 55:9**

As far as we are concerned, in the frame of time in which we exist, God has been there forever and there is no need for Him to have had a start. In the time and space dimensions where God exists, which the Bible refers to as heavenly realms, the workings of science are different to how they are on Earth. We cannot apply our classical understanding of the universe to the environment where God lives, and therefore we cannot prove that God exists by demonstrating it in a laboratory. The differences in scientific environments in Heaven and on Earth may explain why we perceive that He is both immortal and everywhere at once.

Relationships and the way of working within the Trinity have changed since the beginning. At the time of creation Jesus was not the Son of God as He is regarded now, although He was always predestined to be so. Jesus became the Son and God almighty became the Father at the point when Jesus was born and baptised on the Earth two thousand years ago. This is prophesied in the Psalms and recorded in the book of Acts:

> *'God has fulfilled this for us their children, in that He has raised up Jesus. As it is also written in the second Psalm: 'You are My Son, today I have begotten You."*
>
> **Psalm 2:7 and Acts 13:33**

When Jesus was baptised, the voice of God the Father announced from Heaven: *'This is My beloved Son, in whom I am well pleased…'* (Matthew 3:17).

Membership of the Holy Spirit was a divisive issue after the time of Christ and it was not until two hundred years after His death that the Christian church finally agreed that the Holy Spirit is a fully-fledged member of the Trinity. From that time until now, the Trinity has had its familiar form of Father, Son and Holy Spirit.

We assume that the Trinity was around before the creation as described in the Bible book of Genesis, since surely a maker has to exist before anything that he or she makes can exist.

As for the functions or roles of the Trinity, God the Father is the commander, Jesus is the executor and the Holy Spirit is the witness and the power. The name Yahweh means God, it encapsulates all the covenants and attributes of God. The Jews use the name Yahweh when speaking, but consider this name to be too holy to be written down, so it is modified to Jehovah for use in texts. In the Christian church, the name Jehovah God is commonly used to mean God the Father.

Attributes of God

The attribute nature of Father God comprises light and love. Each of these components is itself both physical and spiritual, which encapsulates all aspects of truth that include being right and righteousness.

In terms of physical characteristics, the behaviour of light is well understood and we look at this in chapter 7 on the universe. The physical concept of love, such as how it travels and behaves, is difficult to think about, so we won't try. It is something we feel rather than something we can see and measure.

The spiritual characteristics of light are its beauty, radiance, and utility. We see objects because of the light reflected by them. We know that light emanates from God through His Word to illuminate truth and to dispel darkness. Personalising this, we all tend to harbour thoughts and habits that we do not want exposed to other people or to God, but His light will overcome these areas of darkness if we allow it to. Jesus carries this light to us. In the Bible, John expresses it this way:

> *'In Him [Jesus] was life, and the life was the light of men. And the light shines in the darkness, and the darkness did not comprehend it'.*
>
> **John 1:4-5**

Light is always good for us, but we all know that sometimes it is chastening and unpleasant if we are ashamed of something and would rather it remains hidden. God's other central attribute, love, accompanies any such illumination so that it is carried out without condemnation.

The spiritual love component of God's attribute nature is a love that wants to care, share and encourage; in this respect He is the ideal role model of a parent. Therefore His creation had to take into account sustenance and maintenance of life in a form that would be compatible with Himself, which is mankind, whom He made '*in his image*' according to Genesis 1:26. For mankind to be fully sustained, God knew that we would need the right environment. The implication here is that the universe was created for His special creation that is mankind. This is opposite to the thinking of evolutionism and humanism, which is that the universe somehow came into being first and then mankind developed within it by a long series of seemingly random events.

Love is an emotional and spiritual attribute that motivated God into providing the necessary social aspects of our environment as well as the physical aspects. Communication arises between living beings that can think and make decisions from choices, and this leads to the requirement to take responsibility for such decisions. This is a form of personification of God's attributes; such personification results in acknowledging areas of responsibility. Part of this is an understanding of regulations, showing respect to one another and making the right choices, such the choice to obey God. Social laws have to be made for individuals to collaborate corporately, and God's laws or commandments work along these lines. We see this from the Ten Commandments and from a raft of other laws in Exodus 20. If you violate these laws it is referred to as a trespassing type of sin because you are encroaching on someone else's space, property or life.

People are made up of body, soul, and spirit. Animals have only body and soul, since it was into mankind only that God breathed His spirit in Genesis 2:7. When the human sperm meets the egg at conception, a body is formed for the soul and spirit to dwell within. The soul and spirit are so closely linked, even in scripture, that it is often argued they are one and the same. But we are made in God's image (Trinitarian image). The soul contains our personality, attitudes, character, and memories, and is responsible for the way we think and perceive our environment. As we grow, our souls develop through our circumstances and experiences, but it is our spirits that determine how we act and react to the circumstances. Both soul and spirit need discipline and training. God's Word of truth and wisdom, with the help of the Holy Spirit in our lives, brings about fruit such as love, joy, and peace, through the right attitude attributes as explained in Galatians:

> *'But the fruit of the Spirit is love, joy, peace, longsuffering, kindness, goodness, faithfulness, gentleness, self-control. Against such there is no law.'*
>
> **Galatians 5:22-23**

Jesus

Jesus is referred to as Lord and as the Word of God. Creation was performed through Jesus. We know these things because of the testimony of John:

> *'In the beginning the Word already existed, The Word was with God and the Word was God. He existed in the beginning with God. God created everything through Him and nothing was created except through Him. The Word gave life to everything that was created...'*
>
> **John 1:1-4, NIV**

Jesus was born twice; He was born of Mary physically and then born again after He was crucified. This was declared and proven by Jesus's resurrection from the dead:

> *'...Jesus Christ our Lord, who was born of the seed of David according to the flesh, and declared to be the Son of God with power according to the Spirit of holiness, by the resurrection from the dead...'*
>
> **Romans 1:3-4**

It was necessary for Jesus to be born again so He could fulfil His roles as priest and king. Psalm 89 tells us that Jesus fulfills all history as the messianic King descended from the line of David.

Not only was Jesus born again, He was also firstborn of the dead. Being firstborn of the dead means that Jesus was both the first to rise and the first in supremacy. He is the first to rise from the dead and therefore the first of the new creation, of which He is also the sovereign and inaugurator, responsible for policies including admission. As firstborn He is the rightful heir to it all. We Christians have a sure hope that one day we will follow Christ into the resurrection and the new creation. Because we are in Christ we will reign with Him as the firstborn of God, and be joint heirs with Him of all things in Heaven and on Earth.

In order to reach out to us, Jesus became the son of God and also the son of mankind, so forming a bridge or 'way' between God and mankind. We know this because Jesus stated it:

> *'I am the way, the truth, and the life. No one comes to the Father except through Me.'*
>
> **John 14:6**

Jesus also described Himself thus:

> *'...[I am] the one who is the stairway between heaven and earth on which Angels are ascending and descending...'*
>
> **John 1:51 (NLT)**

The teaching of Jesus is also referred to as '*the way*' by the Roman governor in Caesarea in Acts 24:14 and 24:22.

The Holy Spirit

Although the Holy Spirit has always been around, His way of working has changed since Christ was taken up into Heaven after He rose from the dead. In Old Testament times, the Spirit came on people such as some of the leaders, prophets, and priests. After Christ, with the new covenant, the Holy Spirit actually lives within people:

> *'Or do you not know that your body is the temple of the Holy Spirit who is in you, whom you have from God, and you are not your own. For you were bought at a price; therefore glorify God in your body and in your spirit...'*
>
> **1 Corinthians 6:19-20**

The Holy Spirit is now described as:

> *'the guarantee of our inheritance until the redemption of the purchased possession, to the praise of His [Jesus] glory'*
>
> **Ephesians 1:14**

This means that the same Spirit who raised Jesus from the dead is living within believers. This is an awesome thought. The Holy Spirit's mission is described by Jesus in not-so-gentle terms:

> *'When he [the Holy Spirit] comes, he will prove the world to be in the wrong about sin and righteousness and judgment: about sin, because people do not believe in me; about righteousness, because I am going to the Father, where you can see me no longer; and about judgment, because the prince of this world now stands condemned.'*
>
> **John 16:8-11**

From this we deduce that the Holy Spirit has three works; first the sealing of salvation referred to above, second an anointing for ministry, and third a ministry of conviction and revelation to those who do not yet believe.

Angels, Giants and Demons

Angels

The Bible is primarily about God's relationship with His creation and not about angels, and therefore we don't have much information on them. The Bible gives us fragments that we try to stitch together but the story that emerges is far from complete. Because of this, we cannot be certain of how they came about or how they relate to the Trinity and to us. Here we present some ideas and invite you to draw your own conclusions.

Although angels have been around since before the creation of the Earth, they are themselves created beings. We know this because, as we said earlier, everything has been created by God through Jesus:

> *'For by Him all things were created that are in heaven and that are on earth, visible and invisible, whether thrones or dominions or principalities or powers. All*

things were created through Him and for Him. And He is before all things, and in Him all things consist.'

Colossians 1:16 – 17

This is a reflection of an Old Testament scripture, which tells us that all of the heavenly host, including the angels, worship God:

'You alone are the Lord; You have made heaven, the heaven of heavens, with all their host, the earth and everything on it, the seas and all that is in them, and You preserve them all. The host of heaven worships You.'

Nehemiah 9:6

Angels were either created as part of the creation of the heavens as described in Genesis or perhaps at an earlier time. We cannot put a timescale on this, or even imagine what happened because their creation happened before our time started. The book of Jubilees, which is an ancient book from Judaism and is not part of the Bible, tells us that angels were created along with the rest of the heavens on day one of creation[1]. We can imagine an era before the present one when the Trinity was on its own, so it made sense to create something similar to God (the Trinity) to serve and live with Him through future creations, and so angels came about.

Angels have many tasks and appear in various forms. They can look like men:

'So he lifted his eyes and looked, and behold, three men were standing by him; and when he saw them, he ran from the tent door to meet them...'

Genesis 18:2

---

1 Book of Jubilees, chapter 2 verse 2

And they can look like women with wings:

> *'Then I raised my eyes and looked, and there were two women, coming with the wind in their wings; for they had wings like the wings of a stork...'*
>
> **ZECHARIAH 5:9**

According to Psalm 104:4 God *'makes His angels spirits'*. As for their roles, they do a lot for us. They protect us (Psalm 91:11-12), they let us out of prison (Acts 5:17-21), they feed us (1 Kings 19:5), they deliver God's messages to us (Luke 1:11), and they direct us (Acts 8:26-29).

An example of an angel engaged in the process of healing is where one stirs up the water of the pool of Bethesda in Jerusalem at a certain time, and the first person who enters the water is miraculously healed:

> *'For an angel went down at a certain time into the pool and stirred up the water; then whoever stepped in first, after the stirring of the water, was made well of whatever disease he had.'*
>
> **JOHN 5:4**

Angels are *'ministering spirits appointed to aid the heirs of salvation'*, according to Hebrews 1:14. It is humbling to think that they are appointed to serve God and also to help people who have salvation in Christ. The Greek word for angel means 'messenger' and they are certainly used to convey messages as well as perform other tasks, for God and mankind, throughout the Bible. They will be in control of the world in future ages according to Hebrews 2:5.

According to the scriptures, angels are curious about salvation being available to mankind. It is one of the *'things which angels desire to look into'* (1 Peter 1:12). In relation to covenants, the angelic covenants are different to ours, for there is no salvation or opportunity for repentance for those that rebelled.

Angels were deeply involved in the coming of Jesus to the Earth. There were at least three interventions that are commonly acted out in church nativity plays. First, Gabriel appeared to Zechariah in the temple to announce that he would have a son and instructed him to call his son John, who we now know as John the Baptist. Second, six months later, Gabriel was sent to Mary and told her she had been chosen as the mother of Jesus. Third, an angel appeared to shepherds to announce His birth and then was joined by hosts of others.

A number of seemingly different types of heavenly beings have the role of attending to or worshipping God in Heaven. These include the cherubim who are also called the '*four living creatures*' in Ezekiel, and the seraphim that are mentioned in Isaiah. We believe that the cherubim and seraphim are also angels.

Some believe that the twenty-four elders mentioned in Revelation are also angels. This theory has the difficulty that they wear crowns according to Revelation 4:4, which are indicative of overcoming sin. Angels do not wrestle personally with sin, and it is humans and not angels who are promised crowns. We believe that they are humans who support the church.

Point to ponder: Traditionally the twenty-four elders are considered to be the twelve apostles plus the twelve patriarchs. Is this likely, considering that it mixes gates with foundation stones?

There are only three angels named in the canonical Bible, two good ones and an evil one. The good ones are Gabriel who is described as an angel and Michael who is described as an archangel. The evil one is the archangel Lucifer who became known as Satan after he was cast out of Heaven.

In the book of Enoch, which is not part of the accepted Bible, there are more names of angels. Raphael, Michael, Uriel and Gabriel are described as archangels living in the presence of God, so they must be good. Despite only a few names being known, it is clear that there are a large number of good angels that exist because of what Jesus said as recorded by Matthew:

> *'Or do you think that I cannot now pray to My Father, and He will provide Me with more than twelve legions of angels?'*
>
> **Matthew 26:53**

A legion in the Roman army was around eleven thousand in the time of Jesus when Caesar was emperor. The scriptures talk about the fellowship of the mystery of the beginning of ages, which has been hidden in God who created all things through Christ Jesus according to 1 Peter 1:12, and Ephesians 3:9-13. The good angels will learn some of God's truths through us, and will ultimately be ruled by believers when we are raised from the dead in Christ, according to Hebrews 2:5-9.

Good angels will not accept worship for themselves but point it towards God. They serve God and will never alter the Word of God; they are not seekers of their own power or influence. Evil angels do the opposite.

The manifold, abundant, diverse, and varied wisdom of God shall be made known, explained and experienced by and through the church to the angelic principalities and powers in heavenly places, good and bad. This is according to the eternal purpose that He accomplished in Christ Jesus our Lord according to Colossians 1: 15-18. For this reason the evil angels are condemned and the majority of them are bound up and imprisoned within the Earth. We shall say more about this amazing fact later.

Lucifer, sin and Satan

Lucifer was God's top angelic creation. He was an archangel but a step above the others in beauty, wisdom, and ability. As the covering cherub, he was responsible for worship. Then something went wrong that had devastating consequences for him and the rest of creation. Ezekiel describes Lucifer like this:

> *'[you are] the anointed cherub who covers; I established you; You were on the holy mountain of God; You walked back and forth in the midst of fiery stones. You were perfect in your ways from the day you were created, till iniquity was found in you.'*
>
> EZEKIEL 28: 14-15

He had great beauty and wisdom, which were attributes given to him by God. Yet he was the first being to commit sin.

As it was Lucifer that sinned, only God himself could deal with it (otherwise Lucifer or someone under him could have dealt with it). The story continues:

> *'By the abundance of your trading, You became filled with violence within, and you sinned; therefore I cast you as a profane thing out of the mountain of God; and I destroyed you, O covering cherub, from the midst of the fiery stones. Your heart was lifted up because of your beauty; you corrupted your wisdom for the sake of your splendour; I cast you to the ground...'*
>
> EZEKIEL 28:15-17

It is interesting to think about where the sin came from. We have to assume that sin entered from somewhere outside of Heaven because it did not come from anywhere within the creation. Genesis 1:31 assures us that:

> *'...God saw everything that He had made, and indeed it was very good. So the evening and the morning were the sixth day.'*
>
> GENESIS 1:31

Since what God had made was *'very good'*, there could not have been sin within it at day six of creation. We conclude that to ask how sin managed to enter into Heaven and into Lucifer after day six is similar to the question 'What created God', which we said earlier is a question that mankind cannot attempt to answer. We must accept as a fact through faith that sin entered from somewhere, as described in the Scriptures.

We know that God is a holy God and cannot tolerate sin in His presence, so He had no choice but to cast Lucifer out of Heaven and this is when we begin to think of him as Satan; so Satan is the fallen Lucifer. God also devised a plan to deal with sin itself and, just as importantly, the sin nature, which involves Christ and the church as we will see later.

There are two key scriptures that describe this tragic beginning of sin in the creation, which are Isaiah 14:12-18 and Ezekiel 28:12-18. The Isaiah passage describes how Lucifer said in his heart that, among other things, he wanted to raise his throne above the stars of God and to make himself like the Most High. We have already quoted the Ezekiel passage, which describes how Lucifer sinned through *'abundance of trading'* and subsequently became proud of his own beauty and wisdom. Other Bible translations say *'widespread trading'*.

This theme of trading or buying and selling goes right through the Bible, manifesting on the Earth today as global trading. It is ultimately Satan's undoing, and we pick up this theme in chapter 10 of this book on restoration. We believe that Lucifer carried out widespread trading because he wanted to become independent, and he had a love of money or whatever the equivalent was in his environment. His love of money grew greater than his love for God.

There is some debate about exactly when Satan was cast out of Heaven. On day six, we are told that the creation was *'very good'*, and we assume this means that there was no sin within it. So on day six either Satan had not sinned yet or he had sinned but was somewhere outside of creation. From Genesis 3:1 we know that Satan paid a visit to Eve and this must have been on day seven or later, after Eve was formed. So perhaps he was cast down on day seven and immediately went to meet with Eve.

There is a notion that Satan may not have been actually present himself for his meeting with Eve, but took control of a serpent. He may have used an agent that was vulnerable to his cunning, for we see that after the temptation God punishes the animal. However we tend to think that the serpent was actually Satan because of the reference to him as *'that old serpent'* in Revelation 20:2.

After Satan had deceived Eve, God condemned the serpent body as used by Satan to crawl on his belly and eat dust for the rest of his life (see Genesis 3:14). What a humiliating come-down for who was once the beautiful covering cherub who lived in the presence of God.

Then there is the possibility that Satan was cast down much later, at the beginning of this church age, which seems to have some support from John 12:31 that says: *'Now is the judgment of this world; now the ruler of this world will be cast out'.*

We could interpret this scripture as saying that the ruler of this world (Satan) was cast out from Heaven while Jesus was on the Earth two thousand years ago. There appears to be further support for this idea from Luke 10:18 where Jesus says that He saw Satan *'fall like lightning from heaven'.*

Could John 12:31 and Luke 10:18 mean that Satan was cast down at the point in time when Jesus said these things? Although some believe this, it has significant difficulties. First is that Satan had already sinned when he (or the serpent he controlled) met with Eve. So in the four thousand years between him committing sin and being cast down in the time of Jesus he had to be somewhere other than in the presence of God. Second, being cast down in the time of Jesus doesn't align with the story in the book of Job about Satan coming along uninvited to a meeting called by God. When Satan is asked where he has come from, he replies:

> *'From going to and fro on the earth, and from walking back and forth on it.'*
>
> **Job 1:7.**

Our interpretation of this scripture is that Satan was already confined to the Earth and the air, which he clearly regarded as his property, in Old Testament times.

The last option considered is that Satan could have been cast to the Earth much earlier, even before our creation, where the Earth is described as becoming '*without form, and void*' in Genesis 1:2. But here we are in the foothills of gap theories, which have their own problems that we will discuss separately in the next chapter.

It is apparent to us that the casting down of Satan is happening in stages, and it has two stages yet to go. The first stage was passed when he was cast down to the Earth before he met with Eve. Although confined to the Earth and the air right now, Satan has limited access to at least some parts of Heaven, when permitted, as indicated by Job 1:7. He also has the ability to accuse us before God. He is probably under supervision while there and we can be sure that the purity of Heaven is not degraded by his visits.

Then at the time of the future judgement mentioned in John 12:31, Satan is banished totally from Heaven and is cast down further, being restricted to the surface of the Earth, after losing the war in Heaven as recorded in Revelation 12:7. Then after the judgement he is thrown into the Lake of Fire according to Revelation 20:10.

Point to ponder: Satan is cast out progressively, first from Heaven to Earth and its atmosphere, then to the surface of the Earth, and finally to the Pit within the Earth. This has a parallel with the different stages of the raising of Christ but in the opposite direction, from the Pit to the surface of the Earth, then to the air, and then to Heaven.

Looking again at John 12:31 we see that Satan is described as '*ruler of this world*'. In Job 1:7, Satan says he has been spending his time on the Earth. Some Bible versions translate Job 1:7 differently, with a mix of '*in the earth*' and '*on the earth*'. Interestingly it was probably both, as there is evidence that Hell is inside the interior of the Earth and we discuss this later.

So at the present time Satan's domain is the Earth and its atmosphere. He is currently the legal owner because he holds the mandate, or title deeds, for the Earth that he stole from Adam. Originally God gave Adam dominion over the Earth according to Genesis 1:28, but presently Satan has it. We say more about this later, but be assured that this is a temporary situation; the title deeds are returned in future to their rightful owner, which is Jesus as the second Adam. The title deeds are the scroll with the seven seals that we read about in Revelation chapters 5-8.

Satan somehow found a way into the creation and spoiled it; this has been observed as his general strategy or pattern, distorting and corrupting the good things God has created. He is furious about being cast out of Heaven - and his mission is to kill, steal and destroy. We cannot emphasise enough what an evil and despicable entity he is and the hatred he has for the chosen people (the Jews) and for all those who believe in Jesus. The holocaust during World War II is but one example of his uprisings. He hated Adam and Eve because they were made in the image of God and he hates us for the same reason.

God knew Satan would be like this and hence prepared a place for him and his evil angels:

> *'Then He will also say to those on the left hand, depart from Me, you cursed, into the everlasting fire prepared for the devil and his angels.'*
>
> **Matthew 25:41**

From this we can deduce that the eternal fire was prepared for the Devil and his angels, but the souls of people that Satan has managed to keep away from Christ will also be cast into it.

The good news of the gospel is that Jesus came to destroy the works of Satan, according to 1 John 3:8. When Jesus sacrificed His life on the cross, He actually became the curse or became the sin from the original Adam covenant. Jesus came from heavenly glory, dwelt on the Earth, then went down to the lowest (the abyss that is within the Earth) as the curse and then back to the highest Heaven, which was necessary to fulfil all things (see Ephesians 4:8).

God will restore all things to how creation was intended to be before Adam sinned, but it will be at a terrible cost of life. It is only God himself, the Creator, who can finally put things right and He will; restoration will not take place on this Earth, but on a new Earth. The final chapter of this book discusses how this restoration might happen.

Evil angels

According to Revelation 12:4, one-third of the angels in Heaven turned evil and were cast out along with their master, Satan. If we combine this with our recollection that Jesus was able to call upon twelve leagues of angels from Matthew 26:53, it becomes clear that there are at least a few thousand evil angels. We are told that they are '*his angels*' (Satan's) in Revelation 12:7. Satan is their master because he currently has dominion over the Earth.

We are told that they are in prison in 1 Peter 3:19; we believe that most of them are in prison, but a small proportion remain at large. Satan has assigned this remainder to be responsible for regions, places, people and things. They include powers and principalities that have the objective of deceiving nations. We read about such spirits in the book of Ephesians:

> *'...we do not wrestle against flesh and blood but against principalities, against powers, against rulers of the darkness of this age, against spiritual hosts of wickedness in the heavenly places.'*
>
> **Ephesians 6:12**

More information on the evil angels is given in the book of Jubilees. This book tells us that God commanded that nine tenths of the evil angels be imprisoned, but that one tenth could remain with Satan as their master:

> *'And He (God) said: Let the tenth part of them remain before him, and let nine parts descend into the place of condemnation. And we did according to all His words: all the malignant evil ones we bound in the place of condemnation and a tenth part of them we left that they might be subject before Satan on the earth.'*
>
> **Jubilees 10:11-12**

Since the book of Jubilees is not part of the Bible the information should be treated with caution. However it may explain how and why some of the evil spirits are still active. As for the majority, we are not sure when they were bound and put in the place of condemnation; possibly it was at the time of the flood.

A good angel speaks with Daniel about an evil angel after he (Daniel) had prayed. The angel explains that he would have come immediately to Daniel, but he was delayed:

> *'...the prince of the kingdom of Persia withstood me twenty-one days; and behold, Michael, one of the chief princes, came to help me, for I had been left alone there with the kings of Persia...'*
>
> **Daniel 10:13**

This is an example of an evil angel (principality) that is left in charge of Persia by Satan. Another one is the prince of Greece, whom Daniel tells us will come after the prince of Persia. The angel who met with Daniel goes on to say:

> *'And now I must return to fight with the prince of Persia; and when I have gone forth, indeed the prince of Greece will come'*
>
> **Daniel 10:20**

We believe that the evil angels are the 'sons of God' who rebelled, made an evil pact between them when they came to the Earth, and then had children with human women. These are described in the book of Genesis:

> *'there were giants on the earth in those days, and also afterwards, when the sons of God came into the daughters of men and they bore sons to them. They were the heroes of old, men of renown'.*
>
> **Genesis 6:4**

The Hebrew words here actually mean 'sons of the God'. The same words, with the definite article 'the', are also used for angels in the book of Job:

> *'Now there was a day when the sons of God came to present themselves before the Lord, and Satan also came among them. And the Lord said to Satan, 'From where do you come?' So Satan answered the Lord and said, 'From going to and fro on the earth, and from walking back and forth on it.''*
>
> **Job 1:6-7**

The words used for 'sons of God' later on in Job are slightly different:

> *'or who laid its [the earth's] cornerstone, when the morning stars sang together,and all the sons of God shouted for joy'*
>
> **Job 38:7**

This is actually 'sons of God' (without the definite article). A small difference maybe, but it may be that 'sons of the God' are evil and 'sons of God' are good.

How the evil angels managed to have children with human women is a mystery that we will gloss over, because we don't have an answer. We don't know if they already had sexual organs or whether they had to grow them in some way.

The book of Enoch gives us more information on these evil angels. Please bear in mind that this is also not a Bible book and therefore should be treated with caution; we include this information for interest but do not vouch for its integrity. It gives a more detailed description of the fall of these angels:

> *'And it came to pass that when the children of men had multiplied that in those days were born unto them beautiful and comely daughters. And the angels, the children of heaven, saw and lusted after them, and said to one another: come let us choose us wives from among the children of men and beget us children.'*
>
> **Book of Enoch VI: 1-22**

Chapters VII and VIII go on to explain how these angels were originally sent to watch over the Earth, hence they are called '*watchers*' by Enoch. The actual word 'watching' is used in the NLT version of the Bible in the Job verses that we quoted earlier:

> *'One day the members of the heavenly court came to present themselves before the Lord and the Accuser, Satan, came along with them. 'Where have you come from?' the Lord asked Satan. Satan answered the Lord 'I have been patrolling the earth, watching everything that's going on''*
>
> **Job 1:6-7, NLT**

---

[2] Used with permission. The Books of Enoch, Joseph Lumpkin, Fifth Estate Publishing

It is highly probable that Satan was not expected and not welcome at this meeting. He was singled out and asked where he had come from even though God knew the answer. Maybe God was humiliating him in front of the other angels.

Instead of restricting themselves to watching the Earth, which is the job they were permitted to do after being cast down, the angels started lusting after human women. The book of Enoch tells us that the evil angels taught humans many aspects of art and science that did not have God's approval, such as how to use cosmetics and how to make weapons. We have collected together several verses:

> *'Araqiel taught the signs of the earth. Armaros taught the resolving of enchantments. Azazel: taught the making of weapons of war. Barqel taught astrology. Ezequeel taught the knowledge of the clouds. Gadreel taught the art of cosmetics. Kokabeel taught the mystery of the Stars. Penemue taught writing. Sariel taught the knowledge of the Moon. Semjaza taught Herbal enchantments. Shamshiel taught the signs of the Sun.'*
>
> **Enoch 1, various verses**[3]

We repeat that the book of Enoch is not in the Bible and the information above should be treated with appropriate caution.

The book of Enoch and the Bible book of Genesis both agree that evil practices increased greatly in the human race following the arrival of the evil angels. After the verse describing the activities of the 'sons of the God', Genesis continues:

> *'Then the Lord saw that the wickedness of man was great in the earth, and that every intent of the thoughts of his heart was only evil continually.'*
>
> **Genesis 6:5**

---

[3] Used with permission. The Books of Enoch, Joseph Lumpkin, Fifth Estate Publishing

Sin within mankind started when Satan deceived Eve, and this led Adam to commit sin in the Garden of Eden. Sin propagated from this point through Adam and Eve's son Cain and into the wider population as it grew, but the sin became very much worse when the evil angels arrived. After Adam sinned, God said several things to Adam, Eve and Satan, and this is one thing that He said to Satan:

> *'And I will put enmity between you and the woman,*
> *and between your seed and her seed...'*
>
> **Genesis 3:15**

From this we understand that enmity would be put between Satan's seed and Eve's seed. We can wonder about the mechanisms that Satan could have used to propagate his seed. We think the most probable mechanism was through the evil angels mating with human women.

Why would Satan want his seed to propagate into the human race? We would like to propose a theory: in mating with human women, the evil angels under the leadership of Satan were trying to pollute the blood-line from Adam to Jesus with non-human DNA. It was essential to God's plan of redemption that the blood-line ran from Adam to Jesus through Noah, Abraham, Isaac and the patriarchs of Israel without corruption or pollution, so that Jesus would be purely human and hence be a suitable sacrifice. Satan knew this and tried very hard to introduce impurities. He failed.

This is why Matthew and Luke give the genealogy of Jesus right back to Adam, to show that Satan had failed to corrupt the blood-line. The lists of names in Matthew and Luke are different until they converge at Zerubbabel. Matthew gives Joseph's side, which is important because Joseph was regarded as the legal father of Jesus; here Matthew shows that the kingship from David to Jesus is an unbroken linear series of father-son. The genealogy in Luke is to show that Jesus is the 'Son of man' with the blood-line traced back to Adam; this is believed by Bible scholars to be an account of Mary's line, even though it starts with Joseph. The reason is that it would have been traditional at the time to name the father. So Jesus is Son of God, Son of man, and king.

The book of Enoch gives us even more information about these evil angels. He writes that they came down to the Earth at Mount Hermon, which is in the northern-most part of Israel. Mount Hermon is the only mountain in the area that has snow on it all year round; it is a ski resort. The source of the Jordan river is here; it flows from Hermon and goes underground, emerges again at Banias Springs in northern Israel and then meanders southwards to the Sea of Galilee. It seems a remarkable coincidence that the landing place of the evil angels, who are the source of corruption, was at the same location as the source of the life-giving Jordan.

Satan has shown many times that he likes to occupy the high places. For example, he took Jesus to the highest point on the temple in Luke 4:9 to offer Him all the kingdoms of the Earth if Jesus would bow down to him.

The leader of the evil angels is named as Semjaza in VI:3 of the Book of Enoch. We can speculate that he either is Satan or has a close relationship with him. Enoch describes a hierarchy among them; the evil angels are grouped into tens with a leader over each group.

The evil angels may also have taught mankind how to extract soft metals from the Earth and how to work them. There is an interesting possible connection here with Moses. When Moses came down from Mount Sinai after speaking with God, he found that the Israelites had made a golden statue of a calf from the earrings and other jewellery they had brought out of Egypt. They were singing and dancing around it. The story goes on:

> *'Then he took the calf which they had made, burned it in the fire, and ground it to powder; and he scattered it on the water and made the children of Israel drink it.'*
>
> **Exodus 32:20**

It is not easy to grind gold to powder; it takes skill and specialist tools because it is a soft and malleable metal. Moses grew up in Egypt, so perhaps this knowledge came from the original teaching of these evil angels.

The Nephilim

According to the scripture in Genesis 6:4, the offspring that resulted from the union of evil angels with human women are described as 'heroes of old, men of renown'. You may think this means they were ancient warriors and worthy of respect. In truth they were noteworthy because of their evil, inherited from their fathers. They are known as the Nephilim.

We have heard another idea of how the Nephilim came about, which is that they descended from Adam's sons Cain and Seth. This theory asserts that Cain is the seed of Satan, and the sons of Cain mingled with the daughters of men who descended from Seth. The Nephilim were born from these unions. Now in Genesis 4:7 we are told that sin was '*crouching at the door*' of Cain, and God tells him he must master it, which he failed to do, but there is no mention of him being the seed of Satan. The theory is based upon 1 John 3:12, which says Cain was '*of the wicked one*', but we think this means he came under the influence of Satan, rather than became Satan's seed. We believe the first explanation to be true, which is that the Nephilim were fathered by the sons of God or evil angels, because it is more strongly supported by scripture and because it provides a plausible reason for most of the evil angels being locked up.

The Nephilim resembled giant humans and were all male. It is interesting to consider that there were no giant females; we will return to this thought shortly when discussing how they re-appeared after the flood. The fossils they left behind give us an idea of their height, which is about two to three times the size of an adult human. An example is shown in figure 1.

Figure 1. An example fossil of a giant human beside a rail truck[4]

Many fossils of giant humans, ranging from skulls and odd bones to complete skeletons, have been found in North America, Europe, the Middle East, and the Far East. There are many photographs and videos that can easily be found on scientific, Christian and archaeological web-sites[5].

The observation that giants existed on Earth is an embarrassment to evolution-believing scientists, because the theory of evolution cannot explain it. Fossils of giants are not publicised and never discussed in conferences, indeed most are destroyed or taken away when discovered. Universities are keen to persuade us that photographs of fossils of giant humans are all bogus. In 2002 there was a competition run by Worth2000 to produce the most convincing archaeological fake pictures, which resulted in some fake pictures of human giants. Maybe this was a ruse to cast doubt on all the evidence.

The fossil in figure 1 shows a complete body and it is quite old, published before the days of image processing, so we believe it is one of the genuine pictures. If evidence of the existence of giants

---

4 Used by permission www.biblebelievers.org.au

5 See for example http://archaeologyexcavations.blogspot.co.uk/2012_06_01_archive.html

really is being suppressed and confused as we are led to believe, it is a scientific scandal and outrage. We hope God will soon enable the truth to be revealed. We have noticed that governments are not supportive of the Bible being accepted as an authentic scientific or historical document. Of course, we assert that it is both.

Apart from the fossils and the scriptures, we have anecdotal evidence of the existence of giants from legends. There are stories from many nations that tell of large and highly unusual creatures.[6]

The Nephilim made life a misery for mankind on the Earth: they consumed all the resources, ate people and each other, probably did not need sleep, mated with human women, and indulged in all kinds of evil. Their offspring carried on producing subsequent generations; some of these were also giants, carrying on trying to pollute the blood-line from Adam towards Jesus. God had two physical reactions to the evil influences of the Nephilim and their further descendants.

First, He limited the lifespan of humans to 120 years:

> *'And the Lord said, 'My Spirit shall not strive with man forever, for he is indeed flesh; yet his days shall be one hundred and twenty years"*
>
> **Genesis 6:3**

This age limit was not applied immediately. The Bible records some people that exceeded 120 years of age after this proclamation; among them were Noah's son Shem who lived to 600 years old and Peleg who had a son at 130 years old. People's lifespans gradually reduced to the limit of 120 years. It is interesting that Moses lived exactly 120 years according to Deuteronomy 34:7.

The person that lived the longest time since official records began was a French woman, Jeanne Calment, who died at 122 years old in 1997. At the time of writing, the oldest living person is 116.

---

6 See for example http://en.wikipedia.org/wiki/List_of_legendary_creatures

Second, He brought the flood. The reason that God brought the flood upon Earth in the time of Noah was to destroy the Nephilim, the evil angels and also mankind who were corrupted:

> *'And the Lord was sorry that He had made man on the earth, and He was grieved in His heart. So the Lord said, 'I will destroy man whom I have created from the face of the earth, both man and beast, creeping thing and birds of the air, for I am sorry that I have made them.' But Noah found grace in the eyes of the Lord...'*
>
> **Genesis 6:5-6**

The book of Genesis then straightaway begins to describe the story of the flood in the verses that follow. Note that, at the end of the scripture above, we are told that Noah was one of the few Old Testament characters to find the grace of God.

It is a strange fact but the flood was, in the end, not sufficient to deal with the sin of mankind nor to rid the Earth of the Nephilim permanently, because both re-appeared as we shall see. In Genesis 6:4 the words '*and also afterwards*' are included, indicating that the Nephilim made a re-appearance after the flood. Despite the flood, Satan was not about to give up on his strategy to pollute the blood-line.

There are many references to giants in the Bible throughout the Old Testament following the flood. Over twenty are named that were around at different times, including Nimrod, Goliath, Agag, Og and his father Ogias. An idea of their size can be obtained from a description of King Og, who had a huge bed (or possibly sarcophagus):

> *'King Og of Bashan was the last survivor of the giant Rephaites; his bed was of iron and was more than thirteen feet long and six feet wide. It can still be seen in the Ammonite city of Rabbah.'*
>
> **Deuteronomy 3:11, NLT**

Og was defeated and killed by Moses in Numbers 21:33-35, which says he was the last of the Rephaites. Even so, he was not the last of the giants.

God must have known that the flood would not put an end to either sin or the giants. Although this may seem like a failure to us, remember that God can always see the bigger picture and does things for the best; we believe His plan at that time was to wipe out the corrupt generations and establish another covenant with Noah, which was the next step in His overall covenant plan. Now Noah was a man who kept himself free from sin and corruption. We are told this in Genesis 6:9, which says that Noah was '*perfect in his generations*'. This is why God chose Noah to build the ark that would preserve him and his family. Out of all the people and giants on the Earth, there were only eight people saved from the flood: Noah and his wife, his three sons Shem, Ham and Japheth and their wives, from whom the entire Earth was re-populated.

Our theory of how the Nephilim made a re-appearance after the flood is that someone aboard the ark carried corrupted blood with the giant's DNA. We have just seen that Noah did not carry the corrupted blood since he was perfect in his generations, so who was it? Our suspicion falls onto the wife of Noah's younger son Ham. Being a woman, she could have carried the DNA without being a giant herself. It cannot have been Ham who carried the corrupted blood for two reasons; first, the scripture does not say he was a giant and, second, one of his brothers was an ancestor of Jesus. Shem is named in the genealogy of Jesus given in Luke 3:36, implying that Noah and his wife and all their sons were clean. However, Ham does not seem to have a healthy fear of God, unlike his two brothers. He was the son who did not cover his father Noah's nakedness on one occasion when Noah had drunk too much and fallen asleep:

> *'And Noah began to be a farmer, and he planted a vineyard. Then he drank of the wine and was drunk, and became uncovered in his tent. And Ham, the*

> *father of Canaan, saw the nakedness of his father, and told his two brothers outside. But Shem and Japheth took a garment, laid it on both their shoulders, and went backward and covered the nakedness of their father. Their faces were turned away, and they did not see their father's nakedness. So Noah awoke from his wine, and knew what his younger son had done to him'*
>
> Genesis 9:20-24

We can extrapolate from this thought to the perhaps unfair idea that Ham had chosen a wife who was not of particularly wholesome descent. Note that this scripture also tells us that Ham is the father of Canaan, so his wife was the mother. It was through this line that the Canaanites came about, who were giants descended through Anak according to Numbers 13:33.

Another theory we have heard for the re-appearance of the Nephilim is that, after the flood, there were repeated later occasions when the sons of God came to daughters of men and new populations of giants were started. We cannot exclude the possibility that such later incursions were made, but there is no support for this from the Bible; so on balance we believe that propagation through Ham's wife is the most likely explanation.

Just after the exodus from Egypt, Joshua and the Israelites met the giant Amalekites who were coming the other way. A battle took place. According to Exodus 7:8-13, Joshua and his army defeated the Amalekites while two assistants held up the arms of Moses until sunset.

When the nation of Israel reached the border of the land they had been promised (today's Palestine and Israel), Moses sent twelve spies into the land to carry out a reconnaissance mission. When they came back, ten of the spies reported that the land was full of giants, and that they couldn't be overcome by the Israelites. The other two spies said the giants could be overcome if the Israelites trusted God. A bunch of grapes they brought back was so large that it took two men to carry it. As a result of the majority negative report from the spies, the Israelites lost their faith in God. This is the reason they spent the next forty years wandering in the wilderness, banished there by God until that entire generation had died out.

> *'There we saw the giants (the descendants of Anak came from the giants); and we were like grasshoppers in our own sight, and so we were in their sight.'*
>
> **Numbers 13:33**

The Hebrew word for giants in the above scripture means Nephilim, and this confirms that they are a re-appearance of the giants first described in Genesis 6:4. This later presence of giants in the promised land was a huge stumbling block to Israel. When Israel finally entered the land they had to drive out the giants, but they did not succeed in driving them out completely. A study of giants in the Bible will show that there were several populations of them including the Amalekites, the Amorites, the Canaanites, the Hitites, and the Philistines. The Israelites encountered them all in battle at various times.

The giants and their descendants were always enemies of Israel, persistent foes to make life as difficult as possible for them, and to try to progress Satan's strategy to pollute the human blood-line. God was able to glorify Himself by defeating these enemies supernaturally, and He also tried to stem the pollution of the blood by preventing the Jews from marrying foreign wives. For example we are told about a message given by God through Ezra:

> *'We have trespassed against our God, and have taken pagan wives from the peoples of the land; yet now there is hope in Israel in spite of this. Now therefore, let us make a covenant with our God to put away all these wives and those who have been born to them,*

> *according to the advice of my master and of those who tremble at the commandment of our God; and let it be done according to the law.'*
>
> EZRA 10:2-3

God is very jealous for His people Israel, and He wouldn't tolerate them worshipping other gods or taking wives from foreign peoples. It is a triumphant accomplishment by God that, despite the best efforts of Satan, Israel drove out most of the giants and their descendants, settled in the land they were promised, and preserved the purity of the human blood-line from Adam to Jesus. Jesus was successfully brought forth through God's covenant people Israel.

Demons

We cannot be certain of where demons come from. They are not spirits of evil angels, because demons possess physical structures such as buildings and bodies, whereas angels do not. We have also noticed that angels always operate as individuals whereas demons can operate in groups. One possible source of demons is the evil trinity:

> *'And I saw three evil spirits that looked like frogs leap from the mouths of the dragon, the beast and the False Prophet. They are demonic spirits who work miracles...'*
>
> REVELATION 17:13-14

The evil trinity produces evil, mocking the heavenly Trinity that produces good. Even though the evil trinity is not yet revealed physically on the Earth, they are operating in the spirit and can produce demons. We say more about the evil trinity later. Another possibility is that demons are the spirits of the dead Nephilim, which were present on the Earth in large numbers. Whatever the source of the demons, it seems that there exists an evil hierarchy, where Satan controls the evil angels who in turn control the demons.

We believe there is a high but limited number of evil angels, most of whom are in prison as we have said, but the number of demons may not be limited if they can be produced at will by the evil trinity. Despite Satan's potentially large evil army, their power and how they are deployed are both limited to what God will allow. We know this because when Satan wants to test Job, he has to request permission. An example of God's reply to such a request is: *'Behold, all that he has is in your power; only do not lay a hand on his person.'* (Job 1:12).

During His ministry on the Earth, Jesus cast out many demons that had taken control of people. Demons recognised (and still do recognise) that Jesus is the Son of God. Mark tells us what one group of them said to Him:

> *'Let us alone! What have we to do with You, Jesus of Nazareth? Did You come to destroy us? I know who You are—the Holy One of God!'*
>
> **Mark 1:24**

Luke 8:31 tells us that on another occasion a group of them '*begged Him [Jesus] that He would not command them to go out into the abyss*'. Jesus allowed them to go into the bodies of pigs instead. From this we deduce that demons know their destiny and that there really is a Hell to shun even for them. They were afraid to be cast into the abyss, which implies that the evil hierarchy is run with a great fear of failure.

Demonic activity is still very much with us. Demons can oppress individuals and communities from the outside, and can also internally possess people. It is a gift from God to be able to discern evil spirits and their objectives. Once discerned there are tools available to deal with them. These tools are prayer and fasting, and the authority of the name of Jesus. Casting out of demons is one of the tasks that Christians are sometimes called upon to perform, along with healing the sick and raising the dead:

> *'And as you go, preach, saying, 'The kingdom of heaven is at hand.' Heal the sick, cleanse the lepers, raise the dead, cast out demons. Freely you have received, freely give…'*
>
> **Matthew 10:7-8**

Gap Theories

Gap theories attempt to bend or interpret the creation account in Genesis in such a way that the presence of angels, and the apparent old age of the universe, are taken into account. We assert that these theories are in line with evolutionary thinking and should not be believed. Jesus never referred to any kind of gap, yet He did refer to the Creator, Adam as the first man, and to Noah and the flood.

We believe that gap theories are a convenient compromise between the humanist, who is not accountable to any authority higher than himself (and hence he is his own god) and the creationist, who is accountable to God. The gap theories are for those who wanted both, which is of course impossible. The first we read of a gap theory is when Dr Thomas Chalmers and George Pember introduced one into their teachings. Thomas Chalmers was a professor at the University of Edinburgh, and he formed the Free Church of Scotland. George Pember was an English theologian who set up the Brethren movement. In 1917 Scofield adopted the teaching into his dispensation Bible. In this way the gap theories became a standard interpretation in some fundamental denominations.

Gap theories are founded upon the first two verses of Genesis chapter 1, repeated here:

> *'In the beginning God created the heavens and the earth. The earth was without form, and void; and darkness was on the face of the deep. And the Spirit of God was hovering over the face of the waters.'*
>
> **GENESIS 1:1-2**

Genesis 1 in Hebrew is one sentence; there are no commas or full stops. A candidate gap is between verses 1 and 2 as follows: *'In the beginning God created the heavens and the earth.* – gap – *The earth was without (or possibly became) form and void…'*

This gap allows for two separate creations of the Earth. The idea is that the first Earth became so sinful that God commanded all beings to enter the abyss (or '*deep*') and the Spirit of God then renewed the Earth while hovering over it. This is supported by those who maintain that nothing God created could have become formless and empty on its own, but came about as a consequence of sin.

A variant of this idea proposes that a long time passed between the Earth being created and the advent of day one. During this time, the Earth was or became '*formless and void*'. In answer to this we assert that creation began on day one, and there was nothing physical earlier than this.

Another version of the gap theory states that there was a first creation as described in Genesis chapter 1, followed by a second creation as described in chapter 2. In this theory, Lucifer ruled the Earth in the first creation. Then iniquity is found to have entered him and God destroyed everything and started again in chapter 2. The fallen Lucifer, or Satan, then took control of a serpent to wreak havoc in the second creation.

We assert that there is no evidence of a delay or of a creation when Lucifer reigned. There is no mention in scripture of a destruction at that time whereas other destructions, such as the flood, are fully described. We believe that Genesis 1 is an account of creation, whereas Genesis 2 tells the story again in more detail and from a different viewpoint. When reading Genesis chapters 1 and 2 it is possible that more than one writer is involved as there are at least two different styles.

Some argue that God was 'reconditioning' rather than creating the heavens, the Earth, and its inhabitants during the six day period of Genesis 1, but nothing in the scriptures say or imply anything about reconditioning. All Jewish rabbis accepted the Genesis description of six days for creation and the significance of the seventh day of rest, which permeated much of their teaching. Such a gap theory also does not fit into scriptural types, symbols or numerics.

Apart from Genesis 1:1-2, two other scriptures that are sometimes used to support a gap theory are Jeremiah 4:23 – 26 and 2 Peter 3:5 – 7. We will now proceed to discuss these. The first one is this:

> *'I beheld the earth, and indeed it was without form, and void; And the heavens, they had no light. I beheld the mountains, and indeed they trembled, and all the hills moved back and forth. I beheld, and indeed there was no man, and all the birds of the heavens had fled. I beheld, and indeed the fruitful land was a wilderness, and all its cities were broken down at the presence of the Lord, by His fierce anger.'*
>
> **Jeremiah 4:23 – 26**

Considering the whole context of Jeremiah chapter 4, from verse 5 onwards, it is clear that verses 23-26 are not referring to an early time before creation (or re-creation) of the Earth, but to the destruction of Judah in the end times. The second scripture says this:

> *'For this they wilfully forget: that by the Word of God the heavens were of old, and the earth standing out of water and in the water, by which the world that then existed perished, being flooded with water. But the heavens and the earth which are now preserved by the same Word, are reserved for fire until the day of judgment and perdition of ungodly men...'*
>
> **2 Peter 3:5 – 7**

This scripture tells us that the Earth was formed under water, which conflicts with the explanation from some scientists that the Earth was formed hot and molten; we return to this point later in the book. The words '*were of old*' in verse 5 refer to the time of creation, whereas the words '*that then existed*' in verse 6 refer to the later time Peter was talking about, which is the time of the flood. These are two separate times. Here Peter is saying that the water stored in the Earth at the time of creation was the same water that came from the '*fountains of the deep*' during the flood at a later time. The context of 2 Peter chapter 3 concerns '*scoffers*' in the end times, saying that disaster will come upon them as it did in the time of the flood. The Earth was once destroyed by water and will be destroyed again by fire. Peter is echoing Jesus's words concerning His promised return:

> *'And as it was in the days of Noah, so it will be also in the days of the Son of Man: they ate, they drank, they married wives, they were given in marriage, until the day that Noah entered the ark, and the flood came and destroyed them all. Likewise as it was also in the days of Lot: They ate, they drank, they bought, they sold, they planted, they built...'*
>
> **Luke 17:26-28**

The truth is that on the sixth day God saw that the creation was very good, just as He intended it to be: new and fresh. We conclude that there is no evidence, either from scripture or from science, to support a gap theory, a separate angelic age, or an age when Lucifer ruled on the Earth before he sinned.

Covenants

What covenants are

God has a covenant relationship with His creation. By His creation, we mean the heavens, the Earth, and all living creatures on the Earth. We will describe what we mean by 'covenant' in some detail shortly; for now think of a covenant as a special relationship that involves much commitment. We will also describe what we mean by 'heavens' in some detail later in the book; for now think of them as being the entire universe but not including the Earth itself.

First though, let us discuss whether there is life anywhere else in the universe, intelligent or otherwise. You may wonder why this is relevant. Well, if any were found, it would destroy the case for an omnipotent God who has a covenant relationship with His people here on Earth. This is why governments are spending vast amounts of money on the search for life on other planets. Probes have been sent to Mars and beyond, huge telescopes have been placed in orbit around the Earth, and radio receiving stations have been built around the world, all to try and detect life and especially intelligent life elsewhere.

People who do not believe in God desperately want to justify their unbelief, and proof of life elsewhere would provide this justification. It would mean that we do not need to be accountable to a higher being; we would become our own gods without fear of judgement. Scientists sometimes use arguments of probability to try to persuade us that there must be life elsewhere. An example of such an argument is this: 'there are probably millions of h-class (habitable) planets in the universe, at least one other besides the Earth must have evolved life by chance'. We believe that such arguments are groundless: life does not come about by chance. Life has not been detected on any other planet, and we are certain that it will never be. We believe that God created life here on Earth, God's covenant is with life here on Earth and Jesus died for us here on Earth.

The overall covenant that God has with His creation contains multiple 'component' covenants that are bound up within it. All the component covenants have their time and place within the overall covenant, and many acted as foundations for others. At the present time, the overall covenant between God and His creation is brought together or embodied in Jesus Christ; it is known as the new covenant or New Testament. This is fundamentally a covenant of grace and it has been in force since the time of Jesus's death and resurrection.

Before Jesus came to live on the Earth two thousand years ago, the overall covenant in place was known as the Old Testament, which contained laws that exposed the debt brought about by disobedience and rebellion from the time of Adam, the first man. Within the Old Testament, grace and mercy came upon those who developed a relationship with God and were able to ask for it, such as Noah and Moses, but the people who qualified or were chosen were few and far between. The situation is different now: within the New Testament grace and mercy are available to everyone.

Note that grace and mercy are slightly different from one another; one is giving and the other is holding back. Grace is giving or forgiving when a person does not deserve it, whereas mercy is holding back punishment when a person does deserve it. What Jesus did was to pay the debt with His own blood on the cross, making grace and mercy available to anyone who believes in Him and has a relationship with Him. Grace did exist as part of the Old Testament law (the earlier covenant), but Jesus enforced it. He also clarified and extended the scope of certain laws, such as those relating to divorce, lusting after women, and taking revenge on your enemies. So the New Testament embraces the Old Testament but with a change in focus, and clarification and extension of some of the laws. At the last supper with His disciples, Jesus described it as *'the new covenant in my blood'*, and it began when Jesus shed His blood on the cross.

As mentioned above, within the overall covenant there are several component covenants that are building blocks within it. We have illustrated this in figure 2.

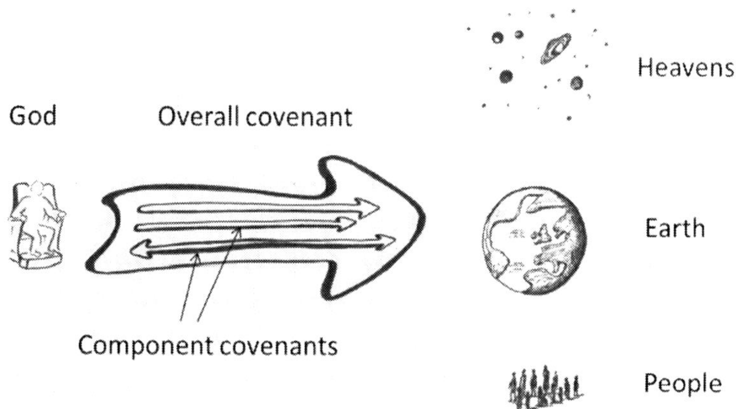

Figure 2. The overall covenant between God and His creation contains several component covenants

Covenants are very strong commitments for aspects of life. They come in two basic types, which are unilateral and bilateral, illustrated in figure 2 by one-way and two-way arrows respectively. We shall give some examples of these types shortly. First though we need to say that covenants have a structure; they consist of the following parts:

» A promise,
» Conditions for bilateral covenants,
» A timescale over which the covenant is in force,
» One or more tokens as reminders of the covenant,
» Description of benefits and penalties (for some covenants),
» Ratification including a blood sacrifice and a meal.

Ratification of a covenant involves the shedding of blood because without blood there is no remission (forgiveness) of sin:

> *'And according to the law almost all things are purified with blood, and without shedding of blood there is no remission...'*
>
> **Hebrews 9:22**

To break a covenant is a very serious matter, it means the invoking of curses or death. A covenant cannot be annulled. Release from a covenant before it has expired can only be achieved through the shedding of blood or through death.

The crux of a covenant is the promise, which is accompanied by agreement of supply, conditions, tokens, the blessings that are benefits of obedience to the covenant, and curses that are the forfeits if the covenant is broken.

A covenant transaction is sometimes called 'cutting the covenant'. Cutting of the covenant would be done at a particular time and place, and in a particular format. An example of literal cutting was the covenant with Abraham, where the token was circumcision or cutting of the foreskin. This means that the seed of Abraham and his descendants flowed through the cut. Circumcision is still practised by orthodox Jews and Muslims who do not embrace the modification of the Abraham covenant by Jesus.

From the point in time when the covenant is cut, you become the covenant holder, and you would personify anything pertaining to the particular covenant taken. You take it on by the way you think, by the way you act, and by the way you live. An example is present day marriage. Man becomes husband, woman becomes wife, and they share the same name and address. In many ways they have both become different people from the time of the covenant onwards; it re-defines them both. Covenant tokens are for recognition and reminders of the covenant. The tokens of the marriage covenant are usually wedding rings, and the timescale of the covenant is until one of the marriage partners dies.

God's covenant of redemption was planned in a way that He would be the instigator, and this allowed him to progressively establish the program of component covenants or, in other words, special individual policies. Eventually, these would all complement each other towards an exhaustive completed plan. This plan covers every possible single detail that could arise in opposition to its ultimate aim and purpose, which is to deal with the sin nature.

Once a covenant is spoken to mankind who hears, believes, and obeys, it becomes established. To enter into a covenant the following steps must be taken: first to hear it, second to accept it, third to believe it, and fourth to act upon it. With the new covenant, Christians do this by daily walking in the covenant 'in Christ'.

In addition to the covenants that are documented in the Bible, it is possible and desirable to have your own personalised covenant. As you walk with Christ, you will encounter various opportunities to establish an 'individual covenant' or policy. It becomes a testimony of God's intervention of His covenant in your life. You have the opportunity of receiving the covenant blessing (victory) but, even if you don't, someone will or someone has, and it is successful from the point of view of the covenant. You would lose out in that case. There will be a judgement day and, on this day, there will be a personified testimony for eternity that will be represented from, through and because of the covenant, which is now all personified in Christ.

At the end of the age, when every spirit and soul that is not covered by the covenant make their defence on judgement day, their excuses are against the covenant and God will call the witnesses forward. Every loop-hole in the covenant must be covered, for two reasons. First, on judgement day when Satan, his evil angels, the demons, and unsaved humans state their defence, every reason and excuse needs to be covered. Second, it is so that eternally saved people know the historical experience of knowledge and wisdom, a kind of reminder. One manifestation of this is the tree of knowledge of good and evil, which is the token of the Adam covenant that was in the Garden of Eden. This was one of two highly significant trees in Eden; the other one was the tree of life.

Both trees will appear again in the new Heaven and the new Earth at the end of the age. Why will the tree of knowledge of good and evil be there, after all the problems it caused in Eden? Well, we need an awareness of the consequences of the sin nature in order to recognise that tree, and we will be reminded of what God has done for us when we are living forever in the new Heaven and new Earth.

Living forever will be possible in complete covenant at that time. Here on Earth when Adam was created, the intention of God was that he would live forever because he had access to the tree of life. However this situation changed after Adam sinned and had eaten from the tree of knowledge of good and evil; he was prevented from eating (or eating further) from the tree of life. From then on, people were prevented from living forever:

> *'Behold, the man has become like one of Us, to know good and evil. And now, lest he put out his hand and take also of the tree of life, and eat, and live forever... therefore the Lord God sent him out of the garden of Eden'*
>
> **Genesis 3: 22-23**

In the book of Revelation, we are told that those who accept salvation and dwell upon the new Earth will once again eat from the tree of life, and live forever.

Let us now look a little more closely at a few of the Old Testament component covenants and how they build one upon another, until we get to the new covenant in Christ and we even get a glimpse of the covenant after that.

Covenant components and completeness

God has an overall covenant with His creation, made up of several component covenants as already mentioned. The Old Testament component covenants are the ones that God made with Adam, Noah, Abraham, Moses, Phineas, and David. The New Covenant and the one after that have Jesus Christ as the mediator. Table 1, over two pages, is a list of these covenants in time order. It shows the covenant holders, whether they are bilateral or unilateral, the promise, conditions, timescales, tokens, consequences of breaking, and ratification meals.

Covenant holder	Unilateral or bilateral	God's promise	Conditions
Adam Genesis 1 and 2	Bilateral	To multiply and fill the Earth and to subdue it Genesis 1:28	Not to eat from the tree of knowledge of good and evil Genesis 2:16
Noah Genesis 6 to 9	Unilateral	Never again to destroy the Earth with water Seed-time and harvest Genesis 8:21-22	
Abraham Genesis 15 to 17	Bilateral	Land and numerous descendants. Shown the new covenant Genesis 15:5	Walk before God and be blameless
Phineas Numbers 25:10-13	Unilateral	Permanent right to priesthood for his descendants, peace	
Israel through Moses Exodus 20 to 25	Bilateral	Land and blessings for Israel	Obey the commandments and the laws
David 2 Samuel 7	Unilateral	Kingdom established for ever	
Christ (new covenant) Gospels	Bilateral	Salvation from Hell and eternal life	Confess that Jesus is Lord and believe that He rose from the dead
Christ Revelation	Unilateral	Married status with Jesus on old Earth and then new Earth	

Table 1, left half. Covenants

Timescale	Token	Consequence of breaking	Ratification meal
Eternal	The tree of knowledge of good and evil	Death and the curse	Fruit and vegetables
Lifetime of the Earth	The rainbow	Not broken	Animal meat Genesis 9:3
Everlasting to descendants	Circumcision Genesis 17:10	Idolatry, slavery and bondage. Destruction of Israel	Bread and wine Genesis 14:18
Everlasting	Not stated	Not broken	Not stated
Everlasting	Two tablets with law written on. Twelve pillars of stone. The Sabbath	Destruction of Israel	A celebration meal before God Exodus 24:11
Everlasting	Eternal Kingship (passed on and given to Jesus)	Not broken	Bread, dates and raisins 1 Chronicles 16:1-3
Lifetime of the heavens and the Earth	Bread and wine	Judgement and assignment of a place in Hell	Bread and wine
Eternal	Future covenant	Cannot be broken	Wedding feast

Table 1, right half. Covenants

The Adam covenant

The first covenant, a bilateral one, was made with Adam and therefore he was the covenant holder. God's promise to Adam (mankind) was this:

> *'...let them [mankind] have dominion over the fish of the sea, over the birds of the air, and over the cattle, over all the earth and over every creeping thing that creeps on the earth'*
>
> GENESIS 1:26

Mankind was instructed to eat a vegetarian diet; eating meat came later with the covenant with Noah.

Later, in Genesis 2:16, God puts Adam into the Garden of Eden, which was then the location of Paradise, to work and take care of the ground. Then God gives Adam further instructions, this time related to Adam's side of the covenant. The covenant condition is this:

> *'Of every tree of the garden you may freely eat, but of the tree of the knowledge of good and evil you shall not eat, for in the day that you eat of it you shall surely die.'*
>
> GENESIS 2:16-17

The tree of knowledge of good and evil in Eden was a token, to remind Adam of the covenant every time he saw it. Now it is worth mentioning at this point that Adam's wife Eve was formed after the terms of this covenant were given. The story of Eve's formation is written in Genesis 2:21-22, which tells us that she was formed from Adam's side. From this we deduce that Adam must have told her about the covenant, because she had some knowledge of it when Satan tempted her. We then read that Adam broke this covenant with God following this temptation of Eve by Satan. In the next chapter we discuss Adam and Eve in more depth, but essentially

Adam broke his covenant by giving away his leadership and by joining Eve in eating from the tree of knowledge.

As a result of Adam breaking his covenant, Satan was able to steal the dominion over the Earth that was given to Adam in Genesis 1:26. In his letter to the Romans, Paul describes this tragedy and explains the consequence:

> *'…death reigned from Adam to Moses, even over those who had not sinned according to the likeness of the transgression of Adam, who is a type of Him who was to come…'*
>
> **Romans 5:14**

The consequences resulting from Adam breaking his covenant were (and still are): enmity between Satan's seed and Eve's seed, painful childbirth for women, marital strife, the soil cursed, presence of thorns and thistles, survival to be a struggle, and death introduced as the inescapable fate of all living things.

The reference in the above scripture to *'a type of Him who was to come'* is a reference to Jesus.

Point to ponder: Jesus came to the Earth as the second Adam and died on the cross two thousand years ago and, in the future, will restore all things to how God originally intended.

The Noah covenant

In the covenant with Noah, mankind's responsibility to populate the Earth was re-affirmed. With this covenant there was a change in the relationship between mankind and animals according to Genesis 9:2. Whereas in the Adam covenant they co-existed in peace with mankind taking care of the animals, from the time of the Noah covenant animals were to live in fear of mankind. This is linked with human diet being expanded to include meat:

> *'Every moving thing that lives shall be food for you. I have given you all things, even as the green herbs. But you shall not eat flesh with its life, that is, its blood.'*
>
> **Genesis 9:3-4**

The Noah covenant was established with all living things aboard the ark:

> *'And as for Me, behold, I establish My covenant with you and with your descendants after you, and with every living creature that is with you: the birds, the cattle, and every beast of the earth with you, of all that go out of the ark, every beast of the earth...'*
>
> **Genesis 9:9-10**

This covenant has a negative implication for the theory of evolution. The implication is that no new species has evolved since the flood, because if one had then it would be outside the Noah covenant, and God's Word would be violated. This is a significant spiritual problem for evolution, but there are many scientific and other spiritual problems that we come to later.

Part of the covenant promise to Noah is that God will never again destroy the surface of the Earth by water. We know from Revelation that the surface of the Earth will be destroyed again, but at that time it will be by fire. The token of the Noah covenant is the rainbow as explained in Genesis 9:13. We still have rainbows, reminding us that this unilateral covenant is still in force.

The killing and eating of animals was permitted from the time of the Noah covenant onwards, and the only exception was not to eat the blood. There is a mix of views about why meat was not allowed to be eaten before the flood whereas it was allowed afterwards. From the covenant point of view, the eating of meat was part of the ratification process.

After the flood, Noah built an altar and sacrificed some of the clean animals upon it. This is one reason that multiple pairs of clean animals were taken on board; according to Genesis 6:20 and 7:2-3, on the ark were seven pairs of all birds, seven pairs of each clean animal and two pairs of each unclean animal.

The Abraham covenant

Abraham started out with the name Abram and his wife Sarah started out with the name Sarai; they both had their names changed by God when this covenant was established.

Abram had a significant meeting with the king of Salem, a high priest named Melchizedek, in the King's valley as described in Genesis 14:17-24. This meeting was significant because Abram gave Melchizedek a tenth of his possessions. Melchizedek blessed Abram, and brought him bread and wine. These tokens of bread and wine are the same ones that Jesus gave to His disciples at the last supper as recorded in the gospels. The implication is that Melchizedek could have been an early appearance of Jesus; the two have other parallels and we explore these shortly.

Later, a heavenly being (probably Jesus again) visited Abram in the form of a smoking oven as described in Genesis 15:17-18, and offered him a bilateral covenant. The oven can be thought of as a type of ark.

In this covenant, God promised Abram that he would possess the land between Wadi and the Euphrates rivers, and that his many descendants would inherit this land. The covenant promise is given in Genesis 15, where God also promised that Abram would have a son and his descendants would be as numerous as the stars in the sky. Abram believed God, and this belief was credited to him as righteousness.

When the heavenly being visited, Abram collected together various animals and cut most of them into two parts, laying the parts side by side on the ground. The ones that Abram did not cut were the birds, because they represent the spiritual aspect; they are of the air. Jesus walked in a figure of 8 around the animal parts according to Genesis 15:17.

The Abraham covenant is everlasting, indeed the blood walk above could be interpreted as the shape of an infinity symbol (∞), and the covenant was ratified by the sprinkling of blood from the collection of animals that Abram had gathered together.

God confirmed this covenant in Genesis 17 after He had renamed Abram to Abraham, and repeated His promise to multiply Abraham's seed so that his descendants would exceed the grains of sand in number. God also promised peace, security, and a land that flowed with milk and honey. The condition of this covenant was that Abraham would walk before God and be blameless.

Now at this point in time Abraham's wife Sarah could not have children, she was barren. This became a matter of some conflict between them, since Sarah was getting on in years and was becoming impatient. She suggested to Abraham that he sleep with her maid Hagar, which he did, and it resulted in Hagar becoming pregnant.

Hagar gave birth to a son, and he was given the name Ishmael. Despite this being Sarah's idea, quite some resentment developed between the two women that resulted in Sarah sending Hagar and Ishmael away. After Hagar departed she had problems finding food and shelter, and an angel visits her with a separate covenant promise:

> *'The Angel of the Lord said to her [Hagar], 'Return to your mistress, and submit yourself under her hand.' Then the Angel of the Lord said to her, 'I will multiply your descendants exceedingly, so that they shall not be counted for multitude...'*
>
> **Genesis 16:9-10**

Hagar was not married to Abraham so she was not in a covenant relationship with him but, because Ishmael was Abraham's son, God was able to bring Ishmael into a covenant, albeit a different one. The

Bible records that Hagar obeyed the angel's instructions; she went back to Sarah and submitted to her.

Abraham prayed to God about Ishmael, to ask if he may live under God's blessing. God replied:

> *'No, Sarah your wife shall bear you a son, and you shall call his name Isaac; I will establish My covenant with him for an everlasting covenant, and with his descendants after him...'*
>
> **Genesis 17:19**

We conclude from this that God promised to greatly multiply the offspring of Hagar's son Ishmael, but the covenant blessing would be upon Isaac, who was yet to be born of Sarah. The 'everlasting' covenant was to be established with Isaac (the son of promise) and not with Ishmael.

It is widely believed that the Arab nations are descended from Ishmael. The Islamic faith was adopted first by these peoples through the work of Mohammed in about 620. Muslims rightly claim that Abraham is their patriarch, through Ishmael. Christians, Jews and Muslims therefore share the same ancestry through Abraham.

Ishmael was born in between God offering the covenant to Abraham and fulfilling it through Isaac. Later in Genesis 16:21, Hagar was sent away for a second time and again an angel visited her; this time the angel told Hagar that Ishmael will be:

> *'a wild donkey of a man; his hand will be against everyone and everyone's hand against him, and he will live in hostility toward all his brothers'*
>
> **Genesis 16:12, NIV**

This is why Islam has always been (and always will be) living in hostility towards everyone who is not a Muslim (so-called infidels), especially Jews and Christians, but more generally towards people from America and the west.

Some fourteen years after Ishmael was born, Abraham had the promised son Isaac with his wife Sarah. It was a long time to wait, but Abraham believed God and trusted in Him to bring about what He had promised.

The reason Islam is particularly hostile to Jews and Christians is that Islam stems from the covenant with Ishmael, whereas both Jews and Christians are beneficiaries of the covenant with Isaac. The conflicts we see between the Arab nations and Israel today have their starting point as the conflict between Hagar and Sarah. Jews have always been heirs of Abraham and, following the sacrifice of Jesus on the cross, Christians are heirs by adoption:

> *'And if you are Christ's, then you are Abraham's seed, and heirs according to the promise'*
>
> **Galatians 3:29**

Let's explore a little more the misunderstanding that Islam has about Abraham and his sons. In Genesis 22:2 the Bible tells us that God told Abraham to take his son to a place called Moriah[7], and to offer the son as a sacrifice. Abraham is obedient and takes his son to Moriah but, as he is about to kill his son with a knife, God says not to do it, but instead to kill a ram that had got caught in a nearby thicket. Now Islam tradition teaches that it was Ishmael who was almost sacrificed at Moriah whereas the Jews and Christians teach that it was Isaac. Verses from the Qu'ran indicate that it was Abraham's 'only' son without naming him. The Bible is crystal clear on the matter; it was Isaac.

The confusion arises because Abraham had two sons, but both the Qu'ran and the Bible state that it was Abraham's only son who

[7] Moriah means 'mountain of the Lord'. The location is believed to be the Temple Mount in Jerusalem

was almost sacrificed. The relevant verse in Genesis, translated from the Hebrew, is this:

> *'Then He said Take now your son your only son Isaac whom you love and go to the land of Moriah and offer him there as a burnt offering on one of the mountains of which I shall tell you.'*
>
> **Genesis 22:2**

We have deliberately left out any punctuation, as this is how the Hebrew is written. Our interpretation of this verse is that Abraham was told to take the only son whom he loved, which was Isaac. The implication here is that Abraham did not love Ishmael, and there is some support from scripture for this. We have seen that in Genesis 16:6 and 21:14, Abraham did not take responsibility for him on two separate occasions. In the first instance Hagar fled with her son. After the angel appeared to her, she returned to Sarah and submitted to her, but then Abraham sent Hagar and Ishmael away again after Isaac was born.

Islamic teaching seems to ignore the fact that Isaac is mentioned by name in Genesis 22:2 above. Isaac was the son of promise, Ishmael was not. It also ignores Genesis 17:19 where Isaac is re-affirmed as the covenant holder following the death of Abraham.

Ishmael was blessed by having one nation descend from him, but Sarah was told that she would become a mother of many nations, and that kings would be descended from her:

> *'...And I will bless her [Sarah] and also give you a son by her; then I will bless her, and she shall be a mother of nations; kings of peoples shall be from her.'*
>
> **Genesis 17:16**

We mentioned that Islam is a hostile religion. Certainly we wish no personal offence to Muslims, but it seems that the Qur'an does not mention God's love explicitly, although there are pointers to it. The Qur'an also does not speak about the forgiveness and redemption of sin through Jesus. It says that Jesus did not die and rise again but was taken directly to Heaven by God. We saw earlier that mercy, grace, and love are all necessary for God's plan to bring mankind back into proper relationship with Himself, which all comes together in Jesus. This means that the grace and love parts are missing from Islam, so that much emphasis is put on memorising the Qur'an and being obedient to it. If you are a Muslim reading this, we implore you to continue to seek out the love and grace of God through Jesus Christ.

Getting back to the story of Isaac, in chapter 3 we said that it was essential to God's plan to maintain the blood-line from Adam to Jesus so that it propagated only through the chosen people Israel. There was a tense moment before Isaac was born when Abraham lent his wife Sarah to Abimelech in Genesis 20:2. Abraham introduced her as his sister, which was a deception but not actually a lie. She was his half-sister; they had the same father. Abraham was afraid that he would be killed if Abimelech knew she was his wife. Sarah was evidently a very attractive woman, despite being at least sixty years old.

God's reaction was to cause Abimelech to have a dream before he had a chance to have sex with Sarah. In the dream God said that she was already married, and told him he would be a dead man if he went ahead. According to Genesis 20:17-18, God also shut all the wombs in the kingdom until Sarah had been returned to Abraham, and Abraham had prayed for Abimelech's restoration. The tense moment passed and Satan's strategy was defeated again; in Genesis chapter 21 we read that Isaac was born to an unpolluted Sarah with Abraham as the father.

In fact this was the second time that Sarah had been believed to be Abraham's sister. The first time, in Genesis 12:10-13, Sarah (then Sarai) told Pharaoh she was Abram's sister following instructions from Abram, because at that time also he was afraid for his life if she told the whole truth. Pharaoh showered Abram with gifts and no doubt had the intention of sleeping with Sarai. God's reaction on that occasion was to send plagues upon Egypt in Genesis 12:17; Pharaoh realised that this was because she was Abram's wife, and sent them on their way unharmed.

We are not told the age of Isaac when he went to Moriah with Abraham, but he was probably not a young boy. When they went, they took two young men, or servants, along with them. The Hebrew word used for the servants is exactly the same word that is translated as '*boy*' in Genesis 22:12, which says '*Do not lay your hand on the boy, nor do anything to him…*'. It is more likely that Isaac was a young man, the same kind of age as a servant.

There are two other indications that Isaac was not a young boy. First, Isaac carried the wood for his own sacrifice, which would only have been physically possible if he was almost adult size. Second, we are told in Genesis 21:34 that Abraham lived many days in the land of the Philistines in between the time that Isaac was born and the journey to Moriah.

Isaac must surely have been a type and symbol for Jesus who was to come. There are several parallels. Isaac and Jesus both carried the wood for their own sacrifice (Jesus carried it part way), they are the only patriarchs who did not have their names changed, and they both spent their whole lives in Canaan. Moreover both Jesus and Isaac were the only sons whom their fathers loved and regarded as suitable for sacrifice. We can speculate that the same thoughts were running through both their minds when they were about to be sacrificed.

There is an even deeper coupling between Isaac, Jesus, and the covenant with Abraham. Abraham's obedience, under the obligatory law of the covenant, actually made it possible and necessary for God to sacrifice His own son Jesus, according to Isaiah 53:4, 6, and 10. God counted Abraham's obedience as righteousness, because it enabled God to proceed with sacrificing Jesus. If Abraham had not been obedient, there would be no salvation now. Just as Isaac was obedient to his father Abraham, so Jesus was obedient to His Father. Just as Isaac was the son of promise of Abraham, so Jesus was the Son of promise of God.

In the garden of Gethsemane while awaiting His arrest, Jesus prayed twice that He would not have to be sacrificed:

> *'O My Father, if it is possible, let this cup pass from Me; nevertheless, not as I will, but as You will… Again, a second time, He went away and prayed, saying, 'O My Father, if this cup cannot pass away from Me unless I drink it, Your will be done"*
>
> **Matthew 26: 39 and 42**

Here Jesus was not being a wimp. He was exploring whether God had a plan similar to the one with Isaac; He had a covenant right to do this. But then after an hour or so of intense prayer, Jesus concluded that He was indeed to be the sacrifice to cover the sins of mankind.

It is significant that Jesus prayed this prayer twice. We believe that, when God repeats a name, such as when He called '*Moses, Moses*' at the burning bush in Exodus 3:4, it is once for the physical realm and once for the spiritual realm. Another example of multiple calling is when God called Samuel three times in 1 Samuel 3:10 for body, soul and spirit. When Paul prayed about his thorn in the flesh in 2 Corinthians 12:7-10, he prayed three times for body, soul and spirit. Jesus replied '*my grace is sufficient for you*' meaning that His grace is sufficient for all things, body, soul and spirit. When Jesus said '*Saul, Saul why are you persecuting me?*' in Acts 9:4, He was arresting Saul physically and arresting Saul spiritually. When He calls just the once, it is to deal with the flesh, the physical or worldly side. Now we have seen that when God spoke to Hagar, He called her just once in Genesis 21:17, to speak about the physical needs of Ishmael. We are not told in scripture explicitly that repeating names is for this purpose, but it is a covenant pattern.

The token of the Abraham covenant was circumcision of males. An important part of this covenant was that those who kept it would be shown the new covenant (in Jesus), but breaking it would lead to the destruction of Israel. The Abraham covenant was broken by Israel's idolatry, which resulted in them being taken into slavery. This would have led to their destruction but for a remedy put in place some four hundred years later, when God developed the Abraham covenant to its next phase.

The Moses covenant

The covenant with Abraham was one on one with God, but the Moses covenant was actually with the whole nation of Israel through Moses. The covenant had several parts as follows:

- » The ten commandments (Deuteronomy 29:1 and 1 Kings 8:9),
- » The law, which is distinct from the commandments, fulfilling the covenant with Abraham (Most of Deuteronomy 29 plus Exodus 19: 5-16),
- » Levitical priesthood (Deuteronomy 18:1-5),
- » The tokens are tablets of stone and the Sabbath when no work should be done; the penalty for disobedience is death (Exodus 31:12-17)
- » Ratification by blood and oath (Zechariah 9:11 and Deuteronomy 29:12),

God gave Moses the law on Mount Sinai (see Exodus 24:1 – 8). One of the tokens is the Sabbath as above, but on Sinai more tokens of this covenant were given. These are the well-known two tablets of stone upon which the law was written by God, but also given were twelve pillars also of stone that represented the twelve children of Jacob (later Israel) and their families, as recorded in Exodus 24:4. The ratification process was the sprinkling of blood on the altar.

Archaeology has revealed that contracts between two nations were sometimes written on two identical tablets of stone, with the same text on each. The stones would be arranged back to back and placed so that the join between them was set upon the boundary between the two nations. The two tablets of stone that God wrote and gave to Moses were identical copies, and can be thought of as symbolically arranged back to back on the boundary between Heaven and Earth, with Heaven and Earth as witnesses to the covenant. This is supported by Deuteronomy 4:26 that says: *'I call heaven and earth to witness against you this day'* (two witnesses are needed).

The twelve pillars of stone were rough long pieces stood on end; the rock represented the established physical Earth, and they were stood on end to represent mankind's part in the Earth. This indicates that we humans have a part in God's mandate for the Earth.

The Israelites side of the Moses covenant was to obey the commandments and other laws that were handed down in Exodus and again in Deuteronomy (see for instance Deuteronomy 5:4 – 20). The Moses covenant bound God to Abraham's seed for 1400 years, after which the covenant was modified yet again. This further modification was required because once again Israel did not keep to its terms.

Although destruction of Israel was the stated penalty for the Moses covenant, God did not do it. When Israel made an idol in the form of a golden calf and began to worship it, God wanted to carry out the destruction, but Moses stood between God and Israel and argued the case with God. He pleaded with God not to destroy them, but to remember that they are His covenant people. Their conversation went like this:

> *'Now therefore, let Me alone, that My wrath may burn hot against them and I may consume them. And I will make of you a great nation.*

Then Moses pleaded with the Lord his God, and said: 'Lord, why does Your wrath burn hot against Your people whom You have brought out of the land of Egypt with great power and with a mighty hand... Turn from Your fierce wrath, and relent from this harm to Your people... So the Lord relented from the harm which He said He would do to His people.'

Exodus 32: 10-14

Moses was righteous enough (and bold enough) to stand like this within the covenant, and plead for Israel. The key words in the above scripture are '*Your people*'. Like Abraham and Noah before him, Moses understood the early symbolic stages of grace under law.

The new covenant

The Moses covenant proved to be impossible for Israel to keep. The difficulty arose because of the need to obey all the law, all of the time. This was not possible because of the sin nature within mankind. God got weary of the continuing animal sacrifices that were being made when people's hearts were not sincere. It was necessary to deal with the sin nature as well as the sin, so a further modification or development was made to the covenant. This was a change whereby the sacrificial law would be fulfilled with Jesus being put to death on the cross, so that the price would be paid once for all time, and redemption completed.

This change ushered in the next phase of God's overall covenant, the New Testament or new covenant, which is a covenant of grace. In some ways this new covenant is even more challenging than the old covenant of law, but along with it comes additional help. This additional help comes from the Holy Spirit or Comforter. He is a member of the Trinity and He actually lives within believers in order to provide the help, and He will continue to do so for the entire time that the new covenant is in force. The end of the new covenant will be signalled by what is called the rapture, and we say more about this later.

The reign of the covenant of grace does not mean that the law has become irrelevant. In Galatians chapter 5, Paul explains that the law is necessary because it defines sin, throws it into relief, but he says not to put effort into obeying the law for its own sake, but to put effort into walking in and being led by the Holy Spirit. In addition, Paul points out that salvation is provided for gentiles (non-Jews) as well as for Israel (the Jews). Moreover, salvation for the gentiles was made possible through the Jews' transgression:

> *'I say then, have they [the Jews] stumbled that they should fall? Certainly not! But through their fall, to provoke them to jealousy, salvation has come to the Gentiles.'*
>
> **Romans 11:11**

From this scripture we can see that God's plan all along was to bring salvation to the gentiles; the mechanism was to allow transgression of the Jews, but not to the point where recovery would not be possible.

Jesus addresses the coupling between the old and new covenants multiple times. He explains that the Old Testament law remains in force until all things are accomplished:

> *'For assuredly, I say to you, till heaven and earth pass away, one jot or one tittle will by no means pass from the law till all is fulfilled.'*
>
> **Matthew 5:18**

Also Jesus makes this amazing statement that tells us He was around before Abraham:

> *'Jesus said to them, 'Most assuredly, I say to you, before Abraham was, I am."*
>
> **John 8:58**

The event known as the transfiguration is interesting to mention here, because it involves three men that represent mankind in the intersection of three phases of the overall covenant. The transfiguration is described in three of the four accounts of the gospel, see for instance Mark 9:2-8. Here, Moses is representing the law and Word (he brought the law and tabernacles), Elijah is representing prophecy and Word (he brought miracles and faith), and Jesus is representing grace and the fulfilled personified Word.

When Jesus died on the cross, it was a sufficient sacrifice to satisfy the sacrificial law, and it changed the focus of the covenant with Abraham that was based on law. As we have seen, Jesus did not remove any of the law, but He fundamentally altered our relationship with it. He removed obedience to the law as a condition for salvation and replaced it with the condition that we confess the lordship of Jesus and believe that He is risen from the dead. The law still stands as a blueprint for living, which you will begin to find yourself following almost as a by-product from your relationship with Jesus and having the Holy Spirit within you.

Now when there is a change of law, there is also a change in priesthood as stated in Hebrews 7:12. When Jesus came, there was a change in priesthood from Levis to Melchizedek; Jesus is described as one after the order of Melchizedek (see Hebrews 6:20), who is now understood to be a 'type' that Jesus followed. Jesus is a high priest of this order, rather than Levite, for at least two reasons.

First, as we wrote earlier in the chapter, Melchizedek oversaw Abraham and his descendants, and Abraham gave tithes to him as described in Genesis 14:18-20. This indicates that Melchizedek, king of Salem, was the more senior man even to Abraham. Melchizedek then blessed Abraham. Second, instead of a priesthood of men who die, need to be replaced, and have to be atoned for, Melchizedek lives forever. He was one of the few people to have been taken directly up to Heaven. Melchizedek is described as:

'having no father and no mother, without descent, having neither beginning of days nor end of life.'

HEBREWS 7:3

This curious description of an eternally living king without genealogy leads us to the thought that he may have been an early appearance of Jesus.

The new covenant, which adds to the law but also brings in grace and the Holy Spirit or Comforter to help us, was explained to the disciples at the last supper, and has bread and wine (ordinary food and drink) as tokens. They are the same tokens that Melchizedek shared with Abraham. This scripture is often read out at communion services in churches:

'...on the same night in which He was betrayed took bread; and when He had given thanks, He broke it and said, 'Take, eat; this is My body which is broken for you; do this in remembrance of Me.' In the same manner He also took the cup after supper, saying, 'This cup is the new covenant in My blood. This do, as often as you drink it, in remembrance of Me..."

1 CORINTHIANS 11:23-25

The new covenant was ratified when Jesus shed His blood on the cross. His broken body is symbolised by the breaking of bread, and His blood poured out for us is symbolised by the cup of wine. So bread and wine are tokens of the new covenant, and each meal we have can be a reminder of it. The timescale of the new covenant is to the end of the age.

Point to ponder: at the crucifixion, did Jesus's blood seep through the ground to the place where the Ark of the Covenant had been buried since the time of Solomon, and 'cover' the ark that contained the two stones from the time of Moses, which were the tokens that represented the covenant under law? This is suggested by the following scripture:

> *'Then he shall kill the goat of the sin offering, which is for the people, bring its blood inside the veil, do with that blood as he did with the blood of the bull, and sprinkle it on the mercy seat and before the mercy seat.'*
>
> **Leviticus 16:15**

Or is it the case that the blood of animal sacrifices was intended only for Earthly tabernacles, but Jesus's blood was intended only for heavenly ones?

It is important to realise that there are several groups of covenant people, all with different covenants. Two of these are the Jews and the church. The Jews are the people whom God calls His own; they are the people that God chose as the route to bring forth Jesus, and hence to restore the whole of creation to Himself. His present goal is to establish the Jews in a land that they can call their kingdom. The future plan is for Jesus to inherit the throne of David and the kingdom, and this is a primary reason for God having a Son. The church is not referred to in the Old Testament, but is described as a '*mystery*' in the New Testament in Ephesians 5:32. The church is the vehicle within which the redemptive work of Christ can be established in individuals. The Jews can progress from the first group to the second; indeed the Jews who believe in Christ, the Messianic Jews, take priority in the process of salvation. They are the 'first fruits', and we read about them in Paul's letter to the Romans:

> *'For I am not ashamed of the gospel of Christ, for it is the power of God to salvation for everyone who believes, for the Jew first and also for the Greek (or gentile).'*
>
> **Romans 1:16**

The church can be viewed as a large group of priests and kings, of which the Jewish priests and kings were a type or precedent.

Future covenant

In the last row of table 1, we have included a future eternal covenant in which Christ marries His bride (the church). This future covenant ushers in an eternity of living with God among people. We get glimpses of it from Revelation and Hebrews:

> *'Now I saw a new heaven and a new earth, for the first heaven and the first earth had passed away. Also there was no more sea. Then I, John, saw the holy city, New Jerusalem, coming down out of heaven from God, prepared as a bride adorned for her husband. And I heard a loud voice from heaven saying, 'Behold, the tabernacle [dwelling place] of God is with men, and He will dwell with them, and they shall be His people. God Himself will be with them and be their God. And God will wipe away every tear from their eyes; there shall be no more death, nor sorrow, nor crying. There shall be no more pain, for the former things have passed away.'*
>
> **Revelation 21:1-4**

In this covenant people will be holy and perfect:

> *'This is the new covenant I will make with my people in that day', says the Lord. 'I will put my laws in their hearts and I will write them on their minds. I will never again remember their sins and lawless deeds'.*
>
> **Hebrews 10:16-17**

This is a unilateral covenant, a gift to those who have been saved through their testimony of Jesus Christ.

It may turn out to be more than one future covenant, for we know that the new covenant is altered when Jesus comes back to Earth in the millennial reign and it is altered again at the time of the end.

We repeat that it was mankind and not God who broke every bilateral Old Testament covenant in table 1. God in His mercy has modified or developed the overall covenant repeatedly to save mankind from the consequences. The heavens and the Earth would be destroyed if God broke any covenants from His side, so that our continuing existence depends upon God keeping them. Jesus expressed this when He said:

> *'And it is easier for heaven and earth to pass away than for one tittle of the law to fail.'*
>
> **Luke 16:17**

We should be eternally grateful to God for His faithfulness to His covenants.

Arks and their relationship with covenants

There are various arks throughout the Bible, and we believe that they relate to God's covenants. They are all boxes or enclosures of some kind, and they all play their part in God's plan for mankind. The first one is the altar built for Abel's sacrifice in the book of Genesis, and the last one is the New Jerusalem in the book of Revelation. Some of them are illustrated in figure 3. They all have meanings and some are connected. For example, Noah's ark saved eight people from the flood, who were Noah and his three sons and all of their four wives. Likewise Jesus can be thought of as our 'ark of salvation' away from the storm of the tribulation that will come upon the Earth.

We came across the smoking oven earlier in the chapter as described in Genesis 15:17-18, where a heavenly being offered Abram the bilateral covenant.

Moses was placed in a basket or 'ark' made of bulrushes by his mother Jochebed. As a Hebrew boy he was in danger of being slaughtered along with all other Hebrew boys by Pharaoh when the Israelites were living as slaves in Egypt. She put him into this ark as a baby to save him, and in a divine twist of irony he was picked up and cared for by Pharaoh's daughter.

Another ark associated with Moses is the Ark of the Covenant, which he made. The priests carried the ark 2000 cubits (about 900m) in front of the Israelites whenever they were on the move. This brought them successfully through wars and other obstacles in their path. There are slightly differing accounts of what was inside this ark, but the most it contained was some manna from heaven, the tablets of stone that were given to Moses with the law written on them (the second set), and Aaron's rod. The contents can be thought of as pointing to Jesus; the manna meant that believers need to be sustained daily, the tablets represented the law and the rod meant authority and identity of priesthood. Aaron's rod was budding, which meant that his priesthood was fruitful. The Ark of the Covenant was made from acacia wood and overlaid with gold, where the wood was representative of mankind and the gold was representative of deity or God. Like Jesus, the ark is the fusing of mankind and God. The ark was carried by Levite priests, almost as if God put Himself in a vulnerable position where mankind had to work with Him when it was being carried. We see this too with the church; God has bound Himself through His Word to mankind. Just as the ark was carried in front leading Israel into battle, so Jesus is leading His church.

The Ark of the Covenant was kept in the tabernacle, in a special compartment that was called the holy of holies. The tabernacle was a large tent during the time when Israel was wandering for 40 years in the wilderness. Wherever Israel camped the tabernacle would be erected. Later, the Ark of the Covenant was in a fixed place in the temple built by Solomon at Jerusalem in special room also called the holy of holies.

Another ark is the manger that Jesus was placed in after His birth, which was a feeding box for animals. This manger could be thought of as an ark of last resort, since there was nowhere else for Mary and Joseph to stay. The arks of Noah and Moses were also arks of last resort. The Moses ark held the baby through whom the law came, and the Jesus ark held the baby through whom grace came.

The next ark is the empty tomb, the empty box that the body of Jesus was put into after His death, but even this could not hold Him. The New Jerusalem is the last ark mentioned in the Bible; it is a huge cube with sides measuring 12,000 stadia along each dimension (about 2,200km).

Mankind fashioned all of the arks except the last one; in fact everything that God has ever asked mankind to fashion can be thought of as an ark or related to an ark. Between them the arks provided places of salvation, protection of the law, places where law was fulfilled and new life came, places of sacrifice and payment of debts, and places where God dwells.

Figure 3. Arks associated with covenants

The Origin and Redemption of Sin 6

Adam and Eve

Adam and Eve were the first people on Earth, according to the scriptures. Adam was formed from the dust of the Earth and then, later on, Eve was formed from a piece taken from Adam's side, most commonly thought of as a rib. According to Genesis 2:21-22, God put Adam into a deep sleep, removed the piece from his side, and then closed him up again. It is interesting that the floating (10th) rib is the only bone that replaces itself.

So when Eve was formed, mankind had the first anaesthetic, the first operation and the first plastic surgery. The word used in the Hebrew for rib is Tsela; this is a feminine word, a root word meaning side. It is also used for a curved rib as in a boat, one side of a door, or can be interpreted as a section of sky, an arch, beam, chamber, leaf, plank, or side of a chamber.

In the Genesis account of creation, all of the creatures that sexually reproduce were created both male and female. Mankind was created slightly differently; as described above, mankind was created and then the female was separated. The special feature of Adam was that he was made in the image of God, unlike all the other beings:

> *'Then God said, 'Let Us make man in Our image, according to Our likeness; let them have dominion over the fish of the sea, over the birds of the air, and over the cattle, over all the earth and over every creeping thing that creeps on the earth.' So God created man in His own image; in the image of God He created him; male and female He created them'*
>
> **Genesis 1:26-27**

The word Elohim is used for God in this scripture; it is also used in Genesis 1:1. We said in chapter 2 that Elohim means plural God, and that this word is half male and half female. The word is also used for an entirely female character, the goddess Ashtoreth in 1 Kings 11:5.

We'll try to make some sense of the mix of singular and plural, and of half male and half female. In the first part of the scripture above, the '*us*' makes sense if the plural God (the Trinity) made the decision to create mankind, but then only one of the Trinity subsequently carried out the actual creating. We have seen in chapter 2 that this was Jesus. Notice also that mankind was created as '*him*' but was created male and female as '*them*'.

The half male and half female character of Elohim provides an insight into the mystery of marriage and the church. Mankind was created male and female, and found their expression in Adam and Eve. Before Eve was formed, Adam had both male and female attributes, since Adam was made in God's image. In order to reproduce sexually, it was necessary for Adam to have a helper. Eve was formed as a suitable helper and, during this process, the feminine parts of Adam were removed and made into a woman. This is strongly supported by the scripture in Genesis that gives Adam's reaction when he first sees Eve:

> *'This is now bone of my bones and flesh of my flesh; she shall be called 'woman,' for she was taken out of man.'*
>
> **Genesis 2:23**

It also is supported by scripture telling us that a man and his wife become '*one flesh*' or one person, in covenant with God, when they marry:

> *'But from the beginning of the creation, God made them male and female. For this reason a man shall leave his father and mother and be joined to his wife, and the two shall become one flesh; so then they are no longer two, but one flesh'*
>
> **Mark 10: 6-8**

God made them male and female, together in one body at the creation, and for *this* reason a man and his wife will become one flesh. Jesus emphasises this point by repeating that '*they are no longer two, but one flesh.*' The one body has male and female attributes.

The taking of Adam's wife from his side has a strong parallel with the blood and water that flowed when Jesus's side was stabbed with a Roman spear. From His blood and water the church was born, the church being to Christ like Eve was to Adam, an essential helper and bride, one flesh with Christ as the head.

In a marriage the covenant is made with the man, and the woman comes into the covenant by relationship with him. The woman is the fulfilment of the covenant; she is the finisher and not the starter. Just as Adam is the covenant holder and Eve is the completer, the body or bride or church of Christ comes out Christ's side where He is the covenant holder and she is His completer. At the present time the church, His body on Earth, is being kept alive by His blood.

In Genesis 1:28 God blesses mankind, and says to them '*fill the earth and subdue it*'. The King James version of the Bible says '*replenish*'. In old English the word literally meant 'fill' or 'make plenty', as in 'fill the Earth'. People who have doubts about Adam and Eve being the first people often ask the question 'where did all the husbands and wives come from?' In answer, we suggest there would have been marriage of brothers to sisters in the early years. After a while there would have been half-brothers and half-sisters, then cousins, and within five generations a high level of diversity would have been apparent. God allowed this to begin with; bear in mind that the laws against incest were given much later to Moses (in Leviticus 18:6).

Even after Noah's flood, there were four wives to four men, which was enough diversity to start off the population again without too much inter-marrying. Men could have more than one wife and this brought forth step-children. This does not apply to the Christian church though. In the New Testament, wives are referred to always in the singular and a standard for leaders is set in 1 Timothy 3:12: '*deacons should be men with one wife, for experience and wisdom in leadership*'.

The original sin of Adam

Adam and Eve were created perfect and without sin. This situation changed however, when Satan paid a visit to Eve. Just as Lucifer was God's top creation, so Adam was the first or 'top' man. When Adam sinned he was reflecting what had happened when Lucifer sinned in Heaven.

In fact it was Eve who yielded to temptation, but then Adam sinned deliberately. Let us explore this a little; the story is an interesting one and we can learn much from it. In Genesis it is recorded that God gave Adam permission to eat from all the trees in the Garden of Eden except one:

> '*but from the tree of knowledge of good and evil do not eat, for in the day that you eat of it, you will surely die*'.
>
> **Genesis 2:16-17**

One purpose of this tree was to present a test of obedience to Adam and Eve, but before Eve was deceived by Satan they had no desire to eat of it.

The words '*for in the day that you eat of it*' seem peculiar, almost as if God is expecting Adam to eat from that tree. This tends to support the theory that Adam was pre-destined to sin. Now it was after this instruction was given that Eve was formed from Adam's side, and therefore we assume that Adam relayed God's instruction to her. We don't know what Adam told Eve about the tree, but when Eve is asked by Satan '*Has God indeed said, 'You shall not eat of every tree of the garden ?*'', she made a critical error. Let us look at her exact words back to Satan. Eve said this:

> *'We may eat the fruit of the trees of the garden; but of the fruit of the tree which is in the midst of the garden, God has said, 'You shall not eat it, nor shall you touch it, lest you die''*
>
> **Genesis 3:3**

Eve's error was that she added to God's Word, since God said nothing about touching the tree. Possibly also she omitted the word 'surely'. Perhaps Adam wasn't very careful when he relayed God's instructions to her; he may have been a bit patronising and said to Eve something like 'don't eat from that tree, in fact don't even touch it or we will die'.

Eve did not reply to Satan with God's Word accurately; she had added to it and also possibly had taken away from it. It was not necessarily her fault, as she may have been repeating what Adam told her. When Jesus was tempted by Satan thousands of years later with a question also related to food, Jesus replied by quoting God's Word exactly, beginning with:

> *'It is written, 'Man shall not live by bread alone, but by every Word that proceeds from the mouth of God''.*
>
> **Matthew 4:4**

We are warned not to add or take away from God's Word in Deuteronomy 4:2, and again in Revelation 22:19. Arguably Eve managed to do both.

From this we can deduce that we must stick accurately to God's Word when dealing with temptation from Satan, otherwise he gets a foothold. Satan knows the Word of God very accurately, presumably he has been through it with a tooth-comb to try and save his own skin, and to prevent as many as people possible being in covenant with God. The written Word of God has authority over Satan and hence over all his evil angels and the demons. It is reasonable to assume that the Word has to be used accurately to carry this authority.

What happened next was that Satan used his foothold to tell Eve, *'you will not surely die'*, (Genesis 3:4) and this directly contradicted God's Word. Satan knew exactly what God had said to Adam, Eve did not. There is also the question of why Adam did not intervene when he saw the serpent talking to his wife; after all he was with her throughout the entire event according to Genesis 3:6, which says:

> *'So when the woman saw that the tree was good for food, that it was pleasant to the eyes, and a tree desirable to make one wise, she took of its fruit and ate. She also gave to her husband with her, and he ate'.*
>
> **Genesis 3:6**

After his discussion with Eve, Satan may have gone away, leaving the temptation to fester within Eve; temptation of good food, pleasure to the eyes, and greater wisdom. Adam stood beside Eve while she took from the tree, and yet he did nothing to stop her. Then she gave him some of the forbidden fruit to eat, and he ate it, so Adam committed a further sin by joining her. Adam passively stood by and let Eve take the lead in this situation; possibly even she allured him, but we know that he gave away his leadership to Eve. This action fundamentally changed the dynamics within his marriage. He forgot his role given to him by God, which included covering Eve, and instead he came under Eve's covering. It is in this way that Adam, the covenant holder, disobeyed and sinned.

Paul says this in his letter to Timothy:

> *'Let a woman learn in silence with all submission. And I do not permit a woman to teach or to have authority over a man, but to be in silence. For Adam was formed first, then Eve. And Adam was not deceived, but the woman being deceived, fell into transgression.'*
>
> 1 TIMOTHY 2:11-14

This scripture may appear at first sight to contradict our assertion that Adam sinned first. In fact they both sinned. This scripture has caused (and still causes) much controversy over the role of women in the family and the church. Upon that matter, there are several examples where women have taken leadership roles in the Bible. An example is Deborah, who was a judge *'over all Israel'* according to Judges chapter 4, which presumably included her husband Lapidoth. God has used and still does use women to lead, whereas the context of the scripture just quoted is that men should be the leaders. We suggest that the answer to this conundrum is that a woman takes a leadership role when there is not a suitable man available. Controversial though this may be, we would point out that much of the Bible appears to be politically incorrect until we understand that God works through covenants.

There is a lesson here for husbands and wives today. The husband is the covenant holder in the marriage, and it is his responsibility to be the leader and keep the marriage covenant intact. Sorry guys, but we will be held accountable to God for our marriages, no matter how well or badly our wives behave, and we should take this role very seriously. The man is the priest and leader in the marriage. In return, wives are blessed when they submit to his leadership and responsibility.

In a marriage, the woman may tend to resent the man's position as leader and covenant holder, and to try and usurp it. Equally, the man may tend to exercise his authority over the woman according to his will. Both of these tendencies are wrong and should be stoutly resisted. Remember that before Adam broke the covenant, men and women were equal in status as created, according to Genesis 1:27. It was only after the fall that God put wives in the position where husbands would rule over them, because Eve took authority over Adam. Peter puts this another way, describing wives as the weaker partner:

> *'Husbands, likewise, dwell with them [wives] with understanding, giving honour to the wife, as to the weaker vessel, and as being heirs together of the grace of life, that your prayers may not be hindered'*
>
> **1 Peter 3:7**

This scripture says that, once married, the woman becomes the weaker vessel and should be treated with understanding and honour. There may be some ladies who have problems with this, but the scripture is clear. The marriage will receive the maximum blessing only when the dynamics are right. Men take note: this scripture tells us that our prayers will not be heard if we fail to honour our wives.

If Adam and Eve had not yielded to temptation, we can guess that Satan would have returned and tried again at 'an opportune time', just as he promised he would return to Jesus as described in Luke 4:13. But this is only speculation; the fact is that they did sin, and God knew that they would. Before the creation of the Earth, Adam was pre-destined to sin. Likewise, Jesus was pre-destined to become the Son of Man and the Son of God, who would not sin, and would destroy the works of Satan as stated in 1 John 3:8.

Adam was formed from the Earth, and God breathed life into him, transferring to him a part of God Himself (Genesis 2:7). The Hebrew word used for 'breathed' here is ruach, literally 'wind of the spirit'. After Adam sinned, this part of God died. As a result, Adam and Eve lost their spiritual covering and they realised they were naked. From then on, scripture says that mankind was born in sin (Romans 5:12), and this is why there is the need to be born again. God made covenant reconciliation, or atonement, by providing Adam and Eve with a covering of animal skin. This was an atonement and not a full restitution back to the original covenant relationship. This atonement was part of His prophetic promise of complete salvation through a saviour. The saviour would be a man from Adam and also a man from God to complete the covenant. Mankind's side of the covenant had been violated, but God's had not.

To provide the atonement covering for Adam and Eve, God was the first to kill an animal and shed blood; this is the first time an animal was killed, a symbolic token sacrifice when God covered them with the animal skin. God continues to cover the church today with robes of righteousness, garments of praise, garments of salvation and spiritual armour, as described in Ephesians 6:10, Isaiah 61:3 and Isaiah 61:10.

After the fall, Adam and Eve were banned from the Garden of Eden, which was the original Paradise where God also walked.

God established a curse, which was His withdrawal from any dominion activity on the Earth, except for activity that was directly involved in the next part of the covenant. We can imagine that thorns and thistles appeared immediately. Tigers grew large threatening teeth and some snakes became poisonous. Creation became suddenly grotesque.

The other very important aspect of Adam's sin is that it enabled Satan to take the title deeds or mandate for Earth, as we mentioned earlier. Satan was able to steal them because Adam had broken his covenant. At the present time the title deeds are still with Satan, who therefore has dominion over Earth that was originally given to mankind. God's plan is to get the title deeds back, which He will succeed in doing towards the end of the age.

The title deeds or mandate appear in Revelation as the scroll with the seven seals, which only Jesus is able to open because He is the slain Lamb of God (see Revelation 5:1-14). The scroll is the will and testament of God the Creator that is legally given to His son Jesus. At the end of the age when God has back the mandate, and therefore has back the legal ownership of the Earth, He will destroy its surface with fire. A new Heaven and a new Earth will provide a home for those who have accepted Jesus. Satan, his angels, the demons, and the souls of people who refuse to believe in Christ will not be admitted. Their fate is to spend eternity[8] in torment, in the Lake of Fire that is prepared for the Devil and his angels, as described in Revelation 20:10.

Propagation of sin

After the fall Adam and Eve were removed from Eden, and sin began to propagate, starting with their immediate family. They had two sons to begin with, possibly twins, where Abel was born first followed by Cain. Cain was the first murderer as described by Genesis 4:1-12, because he killed Abel. It is remarkable that the first human death was by murder. Abel's blood cried out from the ground, giving us insight to the spiritual side; when the body dies, it is not the end.

Cain killed Abel because he became angry, and the root of his anger was jealousy. We explore this a little in the light of what we said in chapter 3 about Lucifer's sin coming about as a result of his widespread trading as recorded in Ezekiel 28:15. We are told in 1 Timothy 6:10 that '*the love of money is a root of all kinds of evil*'. The '*root*' here is the corruption of Lucifer, because he began to love goods or money more than he loved God.

After Adam and Eve yielded to Satan's temptation, they and all the people that followed were also in danger of loving goods or money more than God. To discourage this from happening, God

[8] Literally ages and ages

introduced the system of offering first fruits, which is the giving or sacrificing the best to God.

The source of Cain's anger was that God had accepted Abel's offering but not his own. Abel's offering was of sheep, this being livestock that had their blood shed. We know from Leviticus 17:11 that life is in the blood. Cain's offering was fruit and vegetables, which are seed-bearing and have seed-bearing life. This distinction is often cited as the reason for the rejection of Cain's offering, but we believe that this reasoning is wrong. It was not a blood issue, because these were offerings and not atonement sacrifices, as explained in Hebrews 11:14. The acceptance or rejection of the offerings in this case was based on attitude: Abel gave thankfully, Cain begrudged it.

This is one of the reasons that believers are instructed in the Bible to give a tenth of what they earn (first fruits), sacrificing it to the work of God. Tithing began when Abraham gave the first tenth of his possessions to Melchizedek, and has continued up to the present time. It is not because God needs the money. It is to prevent our love of money becoming greater than our love for God, a protection mechanism against this type of sin and its consequence that is death.

Another example of the love of money attracting the death penalty happened in the early Christian church. Acts chapter 5 verses 1-11 describes how Ananias and his wife Sapphira sold a property, and then made a pact between them to lie to the apostles about how much money they got for it. In this case it was a needless lie; they had the freedom to give what they wanted to the Apostles. Both Ananias and Sapphira died after the apostles confronted them about their lack of honesty and, as a result, great fear came upon the early church. We should all regard this as a warning bell: death and Hell will overtake us one way or another if our love of money is greater than our love for God.

Getting back to Cain and Abel, notice that Cain's is the first account of anger in the Bible, and we see the Lord still has contact with man outside the Garden of Eden after Adam's fall:

'So the Lord said to Cain, 'Why are you angry? And why has your countenance fallen? If you do well, will you not be accepted? And if you do not do well, sin lies at the door. And its desire is for you, but you should rule over it..."

Genesis 4:6-7

The first spiritual principle we have seen is when Adam disobeyed a direct command, but with Cain we see the second spiritual principle where attitude, emotion, compromise, revenge, and so on, can cause wrong thoughts and feelings against someone else. It started with anger caused by God not receiving his offering, and continued with him plotting and carrying out murder. This thinking and planning part is called iniquity. James describes what happened then and what still happens now:

'But each one is tempted when he is drawn away by his own desires and enticed. Then, when desire has conceived, it gives birth to sin; and sin, when it is full-grown, brings forth death'

James 1:14-15

When the opportunity comes or the 'door opens' for sin, what you have thought about in your mind, which is your soul and spirit where the internal hurt and anguish is, determines the resulting behaviour. It becomes a physical deed that is the sin. In the case of Cain it was against another person, and this is called trespass. You violate their space. Cain was not given a verbal command as far as we know, but God had written the law on his heart. But God did warn Cain before it happened. Cain actually understood that it was wrong, for in Genesis 4:9 God asks Cain '*where is your brother?*'. Of course God knew very well, He wanted to test Cain; perhaps it wasn't too late for repentance at that point. Cain replied *'I do not know. Am I my brother's keeper?'*, which was an attempt at avoidance; then it was certainly too late for repentance.

Throughout the Bible God often asks 'what have you done' and similar questions in a particular way to get a particular response from a person or from mankind, so that He in turn can apply a covenant response. For example in Acts 22:7, Jesus asked Saul: '*Why are you persecuting me?* ', rather than 'why are you persecuting my disciples?', leading to the response that began '*Lord...* '. Unlike Cain, in Saul's case repentance was forthcoming.

Propagation of sin has come right down to us individuals. As we stated earlier in chapter 3, the intervention by the evil angels made the sin more intense. We did not choose to be born into a fallen world, but all of us were anyway. It is Jesus that gives us the way out.

Love and salvation

In a strange way, both sin and the sin nature have been dealt with and yet they are still being dealt with. Jesus sacrificed Himself once and for all to cover our sins and their consequences, past, present, and future; yet sin and the sin nature is still being dealt with in our lives. There would be no more sin if the sin nature were to be overcome, and it can be overcome by being born again of the Holy Spirit. The Holy Spirit or Comforter was made available to us when Christ was raised from the dead and went to be with His Father. Being born of the Spirit is a re-birth, a fresh beginning and a disconnection from the first birth that gave us a sin nature.

Recall the computer that we mentioned in chapter 2 when making the point that, in order to create something, we must be outside of it. Imagine now that the computer is fitted to a machine with tracks or wheels, and this machine can make decisions about where it goes. Imagine now that it has been corrupted, maybe the software has been hacked, and the machine is heading for a bad accident that will destroy it. What would you do? You would naturally chase after it and save it from destruction.

There was once a man who kept turkeys. They were running around outside. A storm was coming, and he wanted to move them into the safety of a shed. The stubborn and stupid creatures would not go into the shed, despite his coaxing and running around behind them. He suddenly had the thought: 'if I could become like one of them I could lead them to safety'. He realised the significance of what he had thought, and then went down on his knees and gave his life to Christ.

God realises that we have been born into a fallen world, and we cannot avoid heading for destruction in Hell unless He intervenes. He wants to save us from this calamity, so He became like one of us in the form of Jesus to guide us to safety, and then He died for us to purchase our salvation. If we believe this, and accept the Holy Spirit into ourselves, the Holy Spirit moves in to be with us on the inside to carry on the guiding. At this point we are born again, of the Holy Spirit, and we are able to put aside the life that is driven by the sinful nature. We become a new creation. This is our only hope of salvation, but still He leaves it to us to decide whether or not we want to accept it.

Jesus has dealt with the work of Satan, and also with sin and its consequences through the finished work on the cross. Individual believers who submit to the lordship of Christ, and recognise our complete unity with Him, can live victoriously over sin. This is the only way to defeat the works of Satan and overcome the sin nature in our lives.

Living for Christ may be costly. In the extreme this may involve dying for our faith, and martyrdom has been shown through the Christian age to be a stimulus for the gospel. Christian living does not necessarily mean physical martyrdom, but it does mean that we must 'die to our sinful self' in order to be saved. Jesus said this:

'For whoever desires to save his life will lose it, but whoever loses his life for My sake and the gospel's will save it'

Mark 8:35

We will not be admitted to the new Earth unless we accept that Jesus took the punishment for our sins on the cross, and they are therefore cancelled. Jesus has paid for the sins of the world, but it takes a decision on our part to accept it. Spiritually you can follow Jesus into death so that your sin nature dies, then you are resurrected with Him as a new creation. There is no need for anyone to go into Hell, but you will if you reject this message.

It is staggering to think that believers in Jesus have been chosen since before the creation of the world, but we are told this in Ephesians 1:4. God knew before the creation of the world who would receive salvation through Jesus. This is because He knows the beginning from the end, so even back then He knew who would make it to the new Earth.

The church currently has the mandate to make Jesus known. The mandate will be completed when it has spread the gospel message to all the nations on Earth:

'And this gospel of the kingdom will be preached in all the world as a witness to all the nations, and then the end will come'

Matthew 24:14

The message is reinforced at the end of the book of Matthew in what is known as the great commission:

'And Jesus came and spoke to them, saying, 'All authority has been given to Me in heaven and on earth. Go therefore and make disciples of all the nations, baptizing them in the name of the Father and of the

> *Son and of the Holy Spirit, teaching them to observe all things that I have commanded you; and lo, I am with you always, even to the end of the age.' Amen'*
>
> **MATTHEW 28:18-20**

It bears repeating that God planned for the sin nature to be dealt with through His church, while the church is on the Earth. This is through the work of the Holy Spirit. We cannot overcome the sin nature by any other means. But we must assist the Holy Spirit in this process, by allowing Him to work in every area of our lives, including those that we'd rather not expose.

It is also worth repeating that Adam was a type that Jesus followed. It was necessary for Jesus to be both the Son of Man (of Adam) and the Son of God in order to accomplish this salvation, for all mankind, from the consequences of sin that are death and Hell.

> *'For the wages of sin is death, but the gift of God is eternal life in Christ Jesus our Lord.'*
>
> **ROMANS 6:23**

God planned that a part of Him would go from the highest to the lowest, to redeem the situation for mankind. He did this by not only becoming a man and dying (in the man Jesus), but by actually becoming the curse.

When Jesus was raised from the dead He went the other way, from the lowest to the highest. In order for us to gain a better understanding of why this was necessary and how He did it, we first need to be clear about the relationship between the heavens, the Earth, and Hell.

The relationship between the heavens, the Earth, and Hell

You may have difficulty believing in the literal existence of Heaven and Hell. As well as asking you to believe that Heaven and Hell are real places, we are now going one step further and asking you to believe that Hell is actually inside the Earth. In suggesting this we fully understand and respect the fact that some people will think we are in the realms of fantasy, or stuck with old traditional ideas. We humbly ask that you continue to bear with us; as you read more, we are confident that it will make increasing sense.

We realise that it is not possible to scientifically prove the existence of either Heaven or Hell; certainly they are not observable in the universe at present. Nevertheless please bear with us while we try to ease this difficulty. Remember that faith is the belief in things that are unseen!

Take a look at figure 4. You can see seven levels; the upper three are the heavens, and the lower four layers are the Earth's surface and below the surface.

The lower six levels represent the physical creation or the 'universe'. These are referred to in the creation story as 'the heavens and the Earth' in Genesis. They can be regarded as six levels of mankind, and they have a parallel with the six levels of mankind's relationship with God[9]. Towards the bottom of figure 4 there are three layers under the surface of the Earth; the uppermost one contains the Grave, the next one down contains both the Pit and Paradise that are separated by a gulf, and the bottom layer contains the Lake of Fire. Paradise is currently empty for reasons we will come to shortly. No human soul can cross the gulf between the Pit and the empty Paradise according to Luke 16:26. The Pit contains cavernous places (the Abyss), and also Tartarus, which is the prison where most of the evil angels are locked up. The Pit is the place of torment in Hell where the wicked dead are now; it is also called the Well of Souls by some. At the bottom of the Pit is the Lake of Fire. The Grave is just under the Earth's surface, while the Pit and the

[9] These six levels are potter-clay, shepherd-sheep, king-servant, friend-friend, father-son (or daughter) and lover-love.

Lake of Fire are shown much deeper within the Earth. The Lake of Fire has been prepared for the Devil and his angels according to Revelation 20:10.

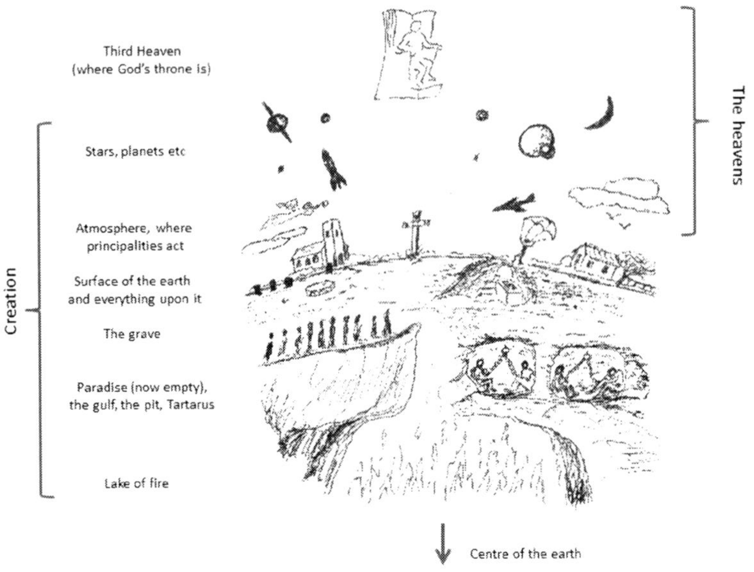

Figure 4. Heaven, Earth, and Hell

At the top of figure 4 is shown the three levels of the heavens. The lowest of them is the Earth's atmosphere and the next one up is the remainder of the known universe. Of course the size of Earth's atmosphere is tiny compared with the universe, so the first heaven is tiny. The third (or highest) Heaven is where God operates from (1 Kings 8:30); we believe that this was present before the creation story in Genesis. The third Heaven is itself sub-divided into different areas or departments in the spiritual realm. Paradise and the throne room are separate areas, and there are probably others. The third Heaven is not directly observable in our dimensions of space and time, but there are objects and projections from our space and time that touch heavenly space and time. We say more about these objects and projections in the next chapter on the universe.

Let us take a look at what physical and scriptural evidence we have for Hell being inside the Earth. Note that Hell does not have to be a physically large space. A spherical space one hundred miles in diameter could accommodate all of the 40 billion people who have ever lived, if their spiritual bodies are the same size as our Earthly ones.

One physical indicator to Hell being inside the Earth is that its interior is simply too hot. The book of Genesis tells us that the Earth was formed under water, which means that the Earth started off very cold and has been heating up since creation. This is contrary to the usual scientific teaching that the Earth was formed hot and molten and has been cooling down ever since.

Currently, the core temperature of the Earth is about 6000 degrees centigrade and the Earth loses heat into space at a rate of about 40 terawatts. So we would expect the core to be getting cooler as time goes by, but it doesn't seem to be; there are even reports on the Internet that it is getting hotter. We are not sure about the trustworthiness of these reports, but if the Earth's core is heating up, this could be responsible for the increased earthquake and volcanic activity we have been witnessing during the past few years.

If we look at the number of earthquakes of all magnitudes, we find that in the 1970s there were around five hundred per year, increasing to two thousand per year by 2000 and to three thousand per year by 2010. However, the number and sensitivity of seismological stations also increased during that time and we would expect to see an upward trend because more of the smaller earthquakes would be detected. So let us look only at the larger earthquakes that have a magnitude of greater than seven on the Richter scale, which would be detectable on any seismometer anywhere.

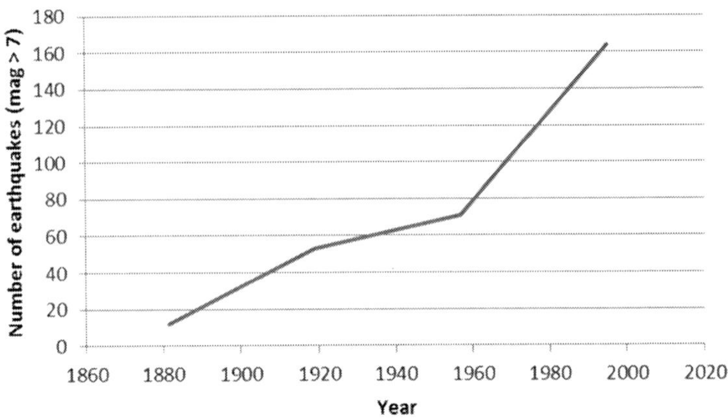

Figure 5. Number of earthquakes of magnitude greater than seven since 1860. The data used in constructing this graph is readily available on the Internet[10]

The Bible has this to say about earthquakes:

> *'There will be famines and earthquakes in various places. All these are the beginning of birth-pains.'*
>
> MATTHEW 24:7-8

If we believe the literal words in Genesis, which tell us that the Earth was made under water, it means that the core temperature has increased from below 100 degrees centigrade (the boiling point of water) to 6000 degrees in six thousand years. An increase in temperature of one degree per year for six thousand years requires an almost impossible amount of heat to be generated inside the Earth from natural processes.

Those who believe the Earth to be 4.5 billion years old also have a problem with the high amount of heat in the Earth at the present time, but for a different reason. If we imagine for a moment that it did form hot and molten all that time ago and has been cooling

10 See for example http://www.Earth.webecs.co.uk/

down ever since, the heat loss through the surface would cause the core temperature to fall to zero degrees centigrade after a few tens of thousands of years. That is, unless there has been an impossibly large amount of heat generated in the Earth to replace that being lost.

So where does the high amount of heat in the Earth's interior come from? It doesn't come from the sun, because not enough heat gets through the Earth's crust to make a significant difference. If you do some research on this topic, you will find that the usual explanation is that a high proportion of the heat (about eighty percent) comes from radio-active processes within the Earth's mantle, and most of the remainder comes from friction caused by movements of the tectonic plates. This is a dubious explanation, because it is far from certain that there is enough radio-active material present inside the Earth to generate this heat. For believers in an old Earth the problem is compounded because radio-active decay falls exponentially with time, and after billions of years would be substantially reduced. For ourselves who believe in a Biblical six thousand year-old Earth, there is also insufficient radio-active heat being generated to account for the necessary core temperature increase of one degree per year.

The conclusion is that we just don't know why the interior of the Earth is so hot. Since science has failed us on this occasion, let us make a suggestion that is radical, from the Bible. The suggestion is that the core of the Earth has been and still is being heated supernaturally, for which there is at least the physical evidence from earthquakes. This thought is rather disturbing, but has an implication that is even more disturbing. God would not be heating up the Earth's interior for any other reason, except that it is the location of Hell.

Now let us consider some scriptures that tell us Hell is within the Earth. Matthew says this:

> *'For as Jonah was three days and three nights in the belly of the great fish, so will the Son of Man be three days and three nights in the heart of the earth.'*
>
> **Matthew 12:40**

This scripture says that it was in the '*heart of the earth*' where Jesus preached to the disobedient according to 1 Peter 3:19. It was necessary for Jesus to spend three days in the lowest part of Hell in order to be able to be raised from the dead. Further scriptural evidence can be found in the Old Testament where we are told how several people disappeared into Hell alive:

> '*...the ground split apart under them, and the earth opened its mouth and swallowed them up, with their households and all the men with Korah, with all their goods. So they and all those with them went down alive into the pit; the earth closed over them, and they perished from among the assembly.*'
>
> **NUMBERS 16:31-33**

More evidence for Hell being inside the Earth is from audio recordings made of human voices crying out in agony deep in the Earth, by workers that were working in mines a long way beneath the surface. These sounds, which terrified the miners, can be readily traced on the Internet [11].

Lastly, there are testimonies from people who have had near-death experiences where they have left their bodies and been taken into Hell for a period of time, and then allowed to return to their bodies. Several people have been taken to Hell for a few minutes by Jesus, and have written books on their experiences. These testimonies and books can be readily identified on the Internet. Some of them have said that somewhere on the Earth's surface is a shaft that leads down into Hell.

So there is a fair bit of evidence that Hell is within the Earth and the volume of this evidence is steadily growing. Jesus is taking more people to Hell for a little while and commanding them to write about their experiences when they return, in order to warn others.

---

[11] For example http://www.youtube.com/watch?v=8iPIXq_jGMQ

All these people have testified that it is an evil place with no love present, only hate, just as the Bible says.

You may wonder about the long term future of people in Hell. It is not pleasant: people who reject Jesus will be thrown into Hell that has been prepared for the Devil and his angels, and stay there forever:

> *'And these [who reject Jesus] will go away into everlasting punishment, but the righteous into eternal life'*
>
> **Matthew 25:46**

The Earth will for ever and ever[12] remain the prison for Satan, the beast and the False Prophet. It will also be the prison for all those who reject Jesus, whose names are not found in the Book of Life:

> *'The devil, who deceived them, was cast into the lake of fire and brimstone where the beast and the False Prophet are. And they will be tormented day and night forever and ever... And anyone not found written in the Book of Life was cast into the lake of fire.'*
>
> **Revelation 20:10 and 20:15**

We think you will agree that eternity is a long time to spend in the wrong place.

Jesus is raised from the dead

We have said that it was necessary for Jesus to go right into the deepest part of Hell and be raised from there, in order to forge the way for our salvation. Now that we have established the relationship between the heavens, the Earth and Hell, we can say more about how He did it. Take a look at figure 6, which shows our interpretation of the layout of Hades, and the path that Jesus took through it.

---

[12] Literally 'ages and ages'

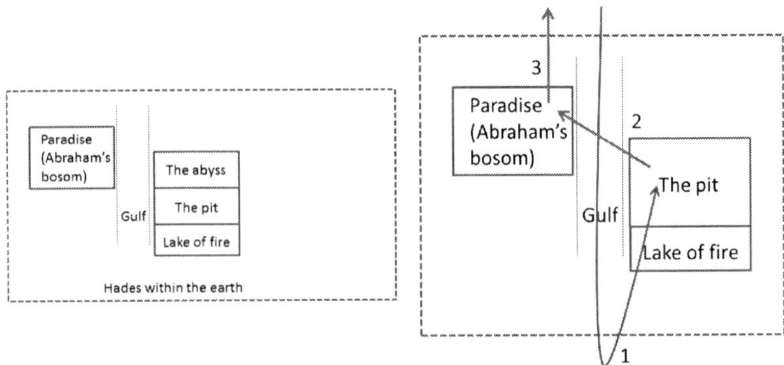

Figure 6. The layout of Hades (left) and the path of Jesus through it (right)

On the first day following His death on the cross, Jesus went into the deepest part of Hell, and came up on the side of the Pit (arrow number 1). He made a proclamation to the demons and the evil angels who are in prison in the Pit and in Tartarus. This was not to help them, but to proclaim judgement and tell them that they had no place in the new covenant. The role of angels is to help humans but the evil angels did the opposite; in effect they had lost their first estate according to Jude 1:6. After He had proclaimed to the spirits and evil angels in prison, Jesus went alone across the gulf to Paradise (arrow number 2 on figure 6).

We know that no human soul can cross this gulf. In the story of Lazarus and the rich man in Luke chapter 16, Lazarus was being comforted by Abraham in Paradise. At that time, Paradise was traditionally called Abraham's bosom. Across the other side of the gulf, they could see the rich man in torment and desperate for water, and the rich man could see them. They could shout at each other, but not get to each other:

> *'And besides all this, between us and you there is a great gulf fixed, so that those who want to pass from here to you cannot, nor can those from there pass to us.'*
>
> **Luke 16:26**

Although no human soul can cross the gulf, we believe that Jesus crossed it and went into Paradise towards the end of day one; we know this because of what Jesus said to the robber who was crucified beside Him: *'today you will be with me in Paradise'*, from Luke 23:43.

In the last section we quoted Matthew 12:40, which tells us that when Jesus died He went into the '*heart of the earth*', and spent three days there. This means that Jesus waited two further days in Paradise and He ascended from there to the surface of the Earth (arrow number 3 on figure 6) after day three. We presume He spent the two days in Paradise talking to the occupants about the new covenant.

At this point it is interesting to briefly consider the question of who raised Jesus from the dead. According to John 2:19, Jesus raised Himself up, whereas Galatians 1:1 says that the Father raised Him up, and Romans 8:11 says that the Holy Spirit raised Him up. So if all these scriptures are to be believed, the entire Trinity was involved. Indeed we believe that this is the case, but they were involved at different stages of Him being raised up. We have said that Jesus actually took on and became the curse. He went into the lowest part of the Earth and, from there, He raised Himself to stage one, which was to the Pit. When He arrived at the Pit, Jesus proclaimed the Word of God to the spirits in prison, and this enabled the Holy Spirit to go and assist Him. Jesus would have used scripture that involved the spirits He was proclaiming to. Now that the Holy Spirit was involved, it was possible for the Father also to become involved either by faith or in actuality. So in this way all of the holy Trinity were involved in raising Jesus from the dead.

Peter refers to Jesus proclaiming to the spirits in prison:

> *'After being made alive, he went and made proclamation to the imprisoned spirits, to those who were disobedient long ago when God waited patiently in the days of Noah while the ark was being built.'*
>
> **1 Peter 3:19-20**

In this scripture the words '*those who were disobedient... while the ark was being built*' indicate that some of the imprisoned spirits are of the Nephilim that God was waiting to kill with the flood. They are presently chained up in the Pit.

When Jesus arrived on the surface of the Earth He appeared to Mary, who did not recognise Him. He instructed her not to touch Him because He had not yet risen to His Father, according to John 20:17. In order to complete redemption for mankind, He had to present His blood to the Father in the perfect tabernacle in Heaven. Allowing anyone to touch Him would have contaminated Him. After this encounter with Mary, Jesus went into Heaven to His Father to present His blood, but then He came back on multiple occasions.

That same day Jesus met two of His disciples on the road to Emmaus in Luke 24:13-35, they also did not recognise Him until He broke bread with them later in the evening, then He vanished from their sight. These meetings with Mary and the disciples indicate that Jesus's appearance had changed somewhat.

At some point around this time, Jesus came back from the Father and went into Paradise within the Earth to collect those who were waiting there with Abraham. He took them up to the new Paradise in Heaven. We know this from Ephesians 4:8, which states that through His grace Jesus '*took captivity captive*'. Jesus has the keys of death and Hades and could (and still can and does) go into Hades whenever He wishes. Death has no hold upon Him. As we said earlier, He accompanies people to Hell for visits. The space that was occupied by Paradise within the Earth is now vacant.

Jesus came back to the Earth on multiple other occasions to visit the disciples; on one such visit He told them to receive the Holy Spirit, made possible because He had gone to the Father. On these later visits the disciples had no problems in recognising Him, implying that his appearance had changed to more resemble His original likeness. Jesus waited a week before paying a visit to Thomas where He invited Thomas to feel the nail-holes in His hands and put a hand into His side, confirming that His body was now able to withstand being touched.

Paradise

It seems that Paradise has moved location twice from the time of creation until now and may possibly move again in future. Before the fall caused by Adam's sin, Paradise was represented by the Garden of Eden. Then, from the time that Adam sinned until the time Jesus died, Paradise was within the Earth where people who died in faith were awaiting the resurrection of Christ. After His resurrection, Jesus moved these people to a heavenly Paradise. Paul confirms that Paradise is currently in Heaven:

> *'We are confident, yes, well pleased rather to be absent from the body and to be present with the Lord.'*
>
> **2 Corinthians 5:8**

This scripture tells us that when believers die, they go home to the Lord who is in Heaven; they do not go to a Paradise that is still within the Earth.

Paradise is a Persian word meaning 'garden to the palace', a place that is planted with flowers and fruit, and has shady places, ponds, and paths. If you had an invitation to see a king, you would be welcomed at the gate and be shown into the 'paradise garden', where you would be told to enjoy the garden and eat the fruit while waiting to be called. It is a waiting place.

The passage of Paradise is illustrated in figure 7.

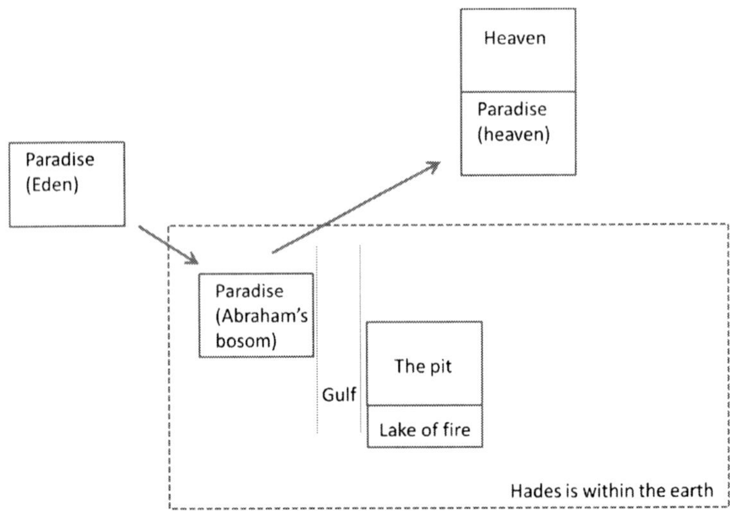

Figure 7. Paradise has moved into and out of Hades in the lifetime of the Earth. The Paradise (Abraham's bosom) part of Hades is now vacant, after the souls in it were taken to Heaven by Jesus

At the present time there are at least two compartments in Heaven, since the location of Paradise is not the same place as the throne of God the Father. Jesus can move in both places though; we know He is in Paradise from what He said to the robber, and we know that He is also in Heaven sitting at the right hand of the Father, from Mark 16:19.

That Jesus can be seen in Paradise in Heaven is also borne out from several accounts of near-death experiences, where people become aware of a bright light, and in the light is Jesus. We know that the radiance from Jesus comes from the Father's glory from Hebrews 1:3, so that these people are in fact seeing the radiance of the Father through Jesus.

Paradise is currently the waiting place in Heaven where the soul of every saved person has been taken and will be taken. At some time in the future, after the restoration, Jesus will present all those who are saved to the Father; then the souls in Paradise will be moved again, this time to the new Heaven and the new Earth.

The Universe

Why the universe was created

In his book 'A Brief History of Time', Stephen Hawking wrote the following words:

> *'Although science may solve the problem of how the universe began, it cannot answer the question: why does the universe bother to exist? Maybe only God can answer that'.*

This great scientist realised that science cannot answer the question of why the universe 'bothers' to exist, and concluded that there is a God who could answer it. We believe we are now in a position to shed some light upon this question.

One reason for creation of the universe (the Earth and the heavens), is that the Earth provides a home for us who God loves. From our home here, we can love Him back. All around the Earth, the heavens can be seen to display His glory. This is described in Psalms 19:1-2.

However an equally important reason is that the Earth is also a platform where sin, the sin nature, and evil spirits can all be dealt with. All these have to be kept out of Heaven. We have seen how Satan has caused mankind to sin, starting by deceiving Eve. God's remedial plan was to bring Jesus to the Earth as a perfect sacrifice. Knowing this, Satan tried his best to corrupt the human blood-line to prevent Jesus being the perfect Son of Man, by introducing his evil angels onto the Earth to mate with human women. This failed and God fulfilled His plan. There needed to be somewhere that is not part of Heaven to fight evil and remove its power. We know that is a process with different stages and we have discussed some of them already. We suggest that if Lucifer had not sinned, we would not be here.

What is the universe ?

Before we can start to discuss the universe any further, we need to be in agreement on what the universe actually is, in the light of the discussion in the previous chapter on Heaven, Hell, and Paradise. Opinions vary on the definition of the universe even within the scientific community, so please try to work with us if our definition of the universe is slightly different from what you think it should be. Throughout this book, what we mean by the universe is 'all matter and the life that is associated with it'. This means the stars, planets and all other physical bodies, and the Earth including everything upon it and within it. It is everything labelled 'the creation' in the lower six levels of figure 4.

The universe is made up of the matter we can observe directly with our eyes, any matter that we can observe indirectly using instruments such as telescopes, and also any matter that we cannot observe directly or indirectly but believe to be 'out there somewhere'. The matter that we think is 'out there somewhere' is not observable either because it is too far away, or doesn't emit light or radio waves so we can't detect it, or because it exists in a higher set of dimensions. We will say more about higher dimensions later.

In some popular science books we have encountered the concept of 'multiple universes' or 'parallel universes'. In our definition these terms have no meaning; we take the view that there is only one universe that contains everything material or physical.

The universe is so large that we cannot see it all. It appears to be infinitely large because we cannot see any boundaries, even with the best scientific instruments. So why does the universe appear to us as it does, infinite and full of stars? We think that it is designed this way to keep us from going insane and also to keep us wondering.

Imagine for a moment that, at night, we could see the edge of the universe, a boundary. Would we not be driven mad by the question of what lies beyond it? Now imagine the opposite, where there are no stars, nothing to see at night. Would we also not be driven mad, because we have no way of judging any kind of distance, and wouldn't know whether the universe was large or small? Surely it is designed to keep us sane, but at the same time to keep us wondering about a Creator, and wondering about how it all works.

Dimensions, time, and expansion of the universe

In this section we give an introduction to the topics of dimensions and time, which will reveal some interesting facts about how the universe is connected to God and Heaven.

Before creation, nothing existed. Physically there was nothing at all, not even space. This is difficult to imagine so let's not try; we just have to accept it in faith, as a fact. Instead let us start by discussing the way that we describe things.

When we try to describe something, we usually say something about its size. For example, the laptop computer I am using to write this is about 300mm wide, 250mm deep, and 25mm high (when the lid is closed). From this, anyone can imagine the size of the object I am talking about. In order to communicate the size I need three numbers: one to give the width, one to give the depth, and the third to give the height. We always need these three dimensions to describe the size of physical objects that we can see with our eyes. It would not be possible to see a two-dimensional object, with the possible exception of a picture, a shadow, or a reflection, but we could argue that these are not really objects in their own right. It is of course possible to describe three-dimensional objects with one or two numbers, rather than three. For example I could describe a football with just one number for its diameter, or a tin of baked beans by two numbers that give its diameter and its length. These are all still three-dimensional objects, but two or more dimensions happen to be the same size.

While discussing dimensions, someone will usually say that a fourth dimension is time, but this is strictly incorrect. Time is measured in units of time, such as seconds, and not in units of length. Students of relativity will use a fourth dimension, which is time multiplied by the speed of light so that the units become length, but this is beyond the scope of the current discussion.

To us, three dimensions are the natural way to describe the size of very tiny objects such as atoms and molecules, right up to very large objects such as stars and galaxies. At the extreme ends of the range of sizes of objects, interesting effects begin to emerge. One such effect is that no matter how small an object is, right down to tiny sub-atomic particles, we can never get to detect the smallest. It is a similar story for the largest object we know, which is the universe; we can't detect the full extent of it. The universe appears to be infinite in size and at the same time infinite in detail. The development of science is not yet adequate to describe it.

Point to ponder: On a logarithmic scale from the size of the known universe (10^{30}m), right down to the smallest objects known about (10^{-30}m), the size of a person is roughly in the centre of the range.

Our common sense tells us that the universe cannot be infinitely large, yet it seems to be. Our common sense also tells us that there cannot be a never ending series of smaller sub-atomic particles, yet there seems to be. Psalm 19:1 tells us that '*The heavens proclaim the glory of God, the skies display His craftsmanship*'. So the scriptures tell us that His glory and craftsmanship are displayed in the universe. The universe surely reflects the infinite greatness of God who created it, and His infinite detail, with mankind in the middle.

We have said that before creation nothing physical existed, not even space. Also before creation, time did not exist. Time did not flow at all. Just like we had to accept that there was nothing at the beginning, not even space, we also have to accept that time did not exist. We are not able to imagine these things, which explains why we would not understand the answer to the question that we discussed at the beginning of chapter 2: Who created God? But despite our human limitations, we believe we can deduce some characteristics of God from what we observe in the universe.

We have seen a few times now the first words in the Bible in Genesis 1:1, which are '*In the beginning, God created the heavens and the earth*', referring to the creation of the universe. Then, Genesis 1:3 describes how time starts: '*then there was morning and evening, the first day*'. So the first two things that God did was to set up the three-dimensional physical system we have been discussing, and set time flowing.

We said earlier that God must exist outside of our dimensions because He created our three dimensions of width, depth and height, and that He could create something only when outside of it. So the fact that He started our time flowing means that God exists outside our time as well as outside our three dimensions. The time flow or clock in Genesis 1 is in relation to the Earth day because it is described in terms of evenings and mornings. This leads to two thoughts about the characteristics of God. First, God appears to us to be everywhere at once because He is outside of our three dimensions; he is omnipresent. Second, God appears to us to have been there forever and will be there forever, because He is outside of our time; He is eternal.

Scientific observations tell us that the universe is expanding. The galaxies in the universe are spreading out; all are moving away from each other[13]. In fact, the further away galaxies are from us, the faster they are moving away.

We said that before creation there was not even space. This means that outside of the universe there is nothing, not even space. The universe is expanding but not into anything. This may be difficult to visualise, but there is a fixed amount of space within the universe and this space is being stretched out as the universe expands. We can imagine that the expansion of the universe is happening a bit like the blowing up of a balloon. Figure 8 tries to illustrate this with a balloon with galaxies printed on it.

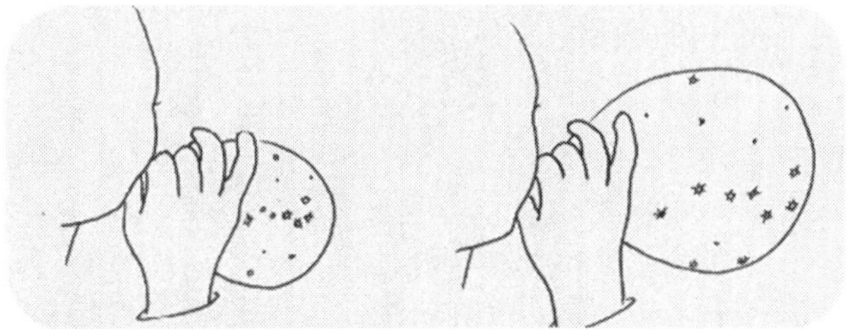

Figure 8. A balloon with galaxies printed on it. When more air is blown in, the balloon expands and the galaxies get bigger and also move away from each other in expanded space

There are seventeen scriptures that tell us God is stretching out the universe; two examples are given here from the Psalms and Isaiah:

> *'...who cover Yourself with light as with a garment, who stretch out the heavens like a curtain'*
>
> **Psalm 104:2**

> *'...I am the Lord, the Maker of all things, who stretches out the heavens...'*
>
> **Isaiah 44:24**

---

13 This is determined from red-shifted light from distant stars. Actually there is one exception, the Andromeda galaxy is slowly approaching our own Milky Way galaxy.

The sky is dark at night

When we look into a cloudless night sky, we see some stars. We can see the closest stars directly with our eyes. If we look with an optical telescope, we see more distant stars. However, beyond a certain distance, we cannot see the stars with our eyes no matter how powerful a telescope we use. The reason is that they have changed colour, into the infra-red and beyond the infra-red so that our eyes cannot see them. The reason the stars have changed colour is that they are moving swiftly away from us; we have said already that the further they are away, the faster they are moving away. In truth the sky is absolutely full of stars. If they were not moving away from us we would see them all, and the night-time sky would be as bright as day-time.

To understand this, we need to consider what is called the Doppler effect, named after its discoverer. If light is being sent from a source (a star) that is moving away from us, the frequency of the light, when received here on Earth, is lower than the frequency when it was sent. As we perceive different frequencies as different colours with our eyes, the colour changes towards red. If the frequency shifts down sufficiently so that the colour moves to the infra-red, our eyes can no longer see the star. We have the same kind of effect with sound; if a police car has its siren running and goes past us at high speed, we hear an increase in pitch as the car approaches, and a decrease in pitch as it goes away into the distance.

Of course we don't notice any change in colour for objects moving on the Earth because they aren't moving fast enough. However every radio engineer knows that if a mobile phone is moving relative to the cellular base-station to which it is connected, the movement causes a slight change of frequency that needs to be corrected by the system, otherwise communication becomes impossible.

In the case of stars, we say that they are 'red-shifted', which means that their light is shifted towards the red end of the light spectrum when they are moving away from us. There aren't any moving quickly towards us, but if there were then we would observe that they would be 'blue-shifted'.

So the sky is dark at night because, although the sky is full of stars, all but the closest ones are red-shifted out of our range of vision. The sky is dark at night because God is stretching out the heavens. Good thing He is, otherwise we wouldn't get much sleep.

What the universe tells us about Heaven

We have established that physical objects such as an everyday box have three dimensions of width, depth, and height, and we mentioned the possible exception of a shadow. Let us now look at shadows in a little more detail. If you shine a torch past an ordinary box so that the light falls onto a screen, a shadow of the box is produced as illustrated in figure 9. A shadow could be described as having two dimensions because it has no depth.

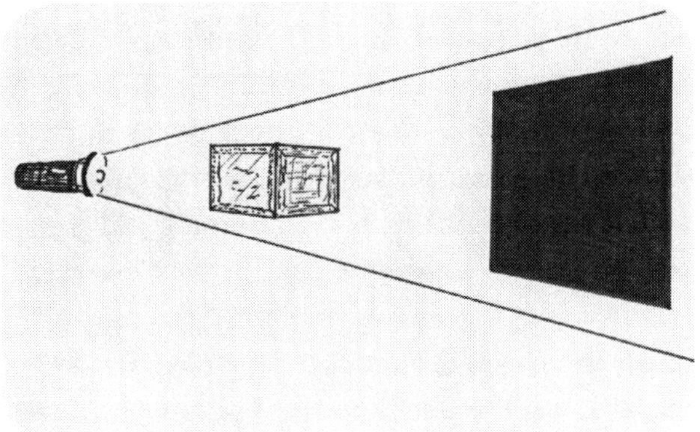

Figure 9. A shadow is cast when light is shone past an object such as a box. The shadow has less information than the object. The shadow is two-dimensional, which is one dimension less than the object that produces it

Our experience tells us that a shadow is a poor representation of the object causing it, with much of the information lost. One of the dimensions is lost because a shadow is a two-dimensional representation of a three-dimensional object. Also, a shadow does not contain any colour information. The box can be very brightly decorated, but this is lost in the shadow.

We would like you to take yet another leap of faith at this point. Imagine that the universe, which is the heavens and the Earth, all that we can see, is a shadow. It is a shadow of the highest Heaven, the shadow being produced or cast by God's radiance.

Just as our experience of a shadow is a two-dimensional representation of an object in three dimensions, we can speculate that the universe is a three-dimensional shadow of something with a greater number of dimensions. The situation is illustrated in figure 10.

Figure 10. The universe could be a shadow of a higher dimensional object (Heaven) illuminated by God's radiance

If the universe is a shadow of the real Heaven, then just like the case where the shadow of a box loses dimensions and detail, the universe loses information from the Heaven that is causing it, including colour, fine detail, and one or more dimensions. People who have had near-death experiences and seen glimpses of the real Heaven tell of how the colours are so much more vivid, and how objects are so much more real and vibrant.

This shadow theory is strongly supported by scriptures; the universe is described as *'displaying God's glory…'* in Psalm 19:1 and also Paul writes *for now we see in a mirror, dimly, but then face to face* in 1 Corinthians 13:12. Another scripture actually uses the word shadow:

'...since there are priests who offer the gifts according to the law; who serve the copy and shadow of the heavenly things'
Hebrews 8:4-5

There is evidence in the Bible that there is a tremendous amount of radiation coming from God the Father. It is beyond what we can physically see or feel or understand, or indeed withstand. Exposure to this radiation would mean instant death. For example Moses was allowed to see the back of God passing by, but only after he was hidden for protection in the cleft of a rock in Exodus 33:22. The radiation from God's face would have killed Moses even though he was protected by solid rock. Can you imagine God's radiance? It is powerful enough to create and sustain the universe as a shadow.

In fact we had already reached the conclusion that God exists in higher dimensions, when we said earlier that God must exist outside our dimensions of space and time in order to create them. It all makes a kind of sense: God's radiance shines through the heavens that are of higher dimensions than the universe, to produce the universe as a shadow in three dimensions.

Having said all this, we must now confess that the universe is more complicated than a three dimensional shadow, even though that is already a fairly complicated concept. The three-dimensional system, although adequate for everyday objects that we can see and pick up, is not adequate to deal with some of the phenomena and objects within the universe. For these we need to consider higher dimensions, but this does not invalidate our theory that Heaven is the real thing, and that the universe is a projection of it. Rather, we can regard the objects that require description by higher dimensions as being connected more intimately with Heaven than some other objects. We look at some of these the next sections.

Things are never simple with God, and the story will never be completely told. This is true of physical objects, the Bible and God Himself. It is a theme that runs through this book: the more you look, the bigger the story becomes, and the more detail there is to see.

Point to ponder: in mathematics there are objects called fractals, which have the property that the deeper you look into them, the more detail you see. God's Word is like this. Matter is like this. Perhaps God Himself has a fractal nature.

Gravity and black holes

The effect of gravity is so common to our experience that we probably don't think about it much. We know that if we drop an object, it falls to the ground. Some force is attracting the object to the ground and we call this force gravity. The strength of the force (or, loosely, the weight) actually depends upon three things: the mass of the object, the mass of the Earth and the height of the object above the Earth.

If we move an object further and further away from the surface of the Earth, the force of gravity that it feels will get weaker and weaker, but it will still be felt. One way of illustrating this is by imagining that the Earth is a marble that is placed on a thin rubber sheet with lines drawn on it and then the sheet is held up by the edges, as shown in figure 11.

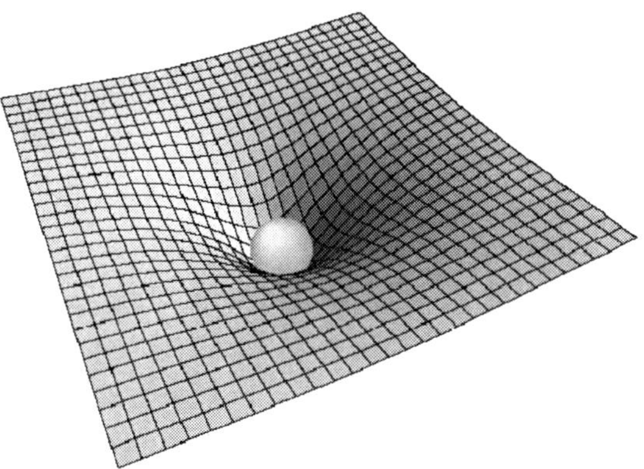

Figure 11. A marble causes an indentation in a rubber sheet, which is representative of the Earth causing an indentation in space[14]

[14] Graphic courtesy of the LIGO Laboratory

The marble makes an indentation in the sheet, and the steepness of the indentation is highest near to the marble. If a smaller object is dropped anywhere onto the sheet, it will tend to roll over to the marble. The smaller object will also produce a smaller distortion of its own that will move the marble slightly. The moon can be thought of as a smaller object that rolls around the indentation caused by the Earth.

In this way we can imagine the Earth 'bending' space so that other objects are attracted to it. Now try to imagine this shape of the bent space occurring all around the Earth, reducing in strength as it gets further away in every direction. You will find that you cannot; it is impossible to imagine this shape in three dimensions. Yet it does happen in reality; gravity is felt in every direction away from the Earth, and the strength diminishes with distance. This is one of the situations that Einstein formulated in his theory of general relativity, and he needed to construct equations in higher dimensions in order to accomplish it. This leads us to conclude that, in the large-scale universe, the action of our common experience of gravity is connected with higher dimensions. This kind of bending in space happens around every object in space; planets, stars, and so on. This is indeed what gravity is, space curving or bending around all objects in the universe, so that other objects are attracted to them[15].

Now there are objects in the universe that have extremely strong amounts of gravity. The most extreme of these are black holes. Black holes are very dense objects, which means that they have an enormous amount of mass compared with their size. To illustrate how dense they are, think of the size of the Earth; it is approximately a sphere of about 5,800km radius, and has a mass of about $6 * 10^{24}$kg (6 with 24 zeros after it). Now imagine that the whole Earth could be compressed into a size of about 2cm radius, about the size of a golf ball. Its density would become enormous. Figure 12 illustrates this.

---

[15] Strictly, both space and time are distorted in the presence of objects, but we do not pursue the time element in this section.

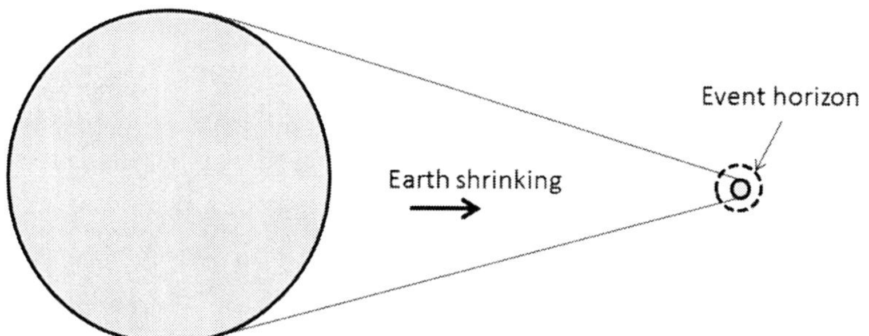

Figure 12. If the Earth were to be compressed so it is smaller than a golf ball, the event horizon would be exposed

The amount of gravity produced by a golf-ball sized Earth would be the same as that produced by the normal sized Earth, because the mass has not changed. But the strength of gravity at the surface at 2cm from the centre would be much higher than that at the current surface at 5,800km, because the same amount of gravity is emanating from a much smaller surface area. The strength of gravity would be so high that even light would be attracted to it, and would not be able to escape. So it would appear to be black.

The size to which an object of any given mass would need to be compressed in order to attract light is called the 'event horizon'. If the size of the object is smaller than its event horizon, so that the event horizon is outside the object, then we have a black hole; so named because not even light can get away from it. The Earth would be a black hole if compressed smaller than a golf ball (2cm radius), and the event horizon would be at 2cm.

If, instead of the Earth, the sun were to be compressed in size in a similar manner, it would not have to shrink to such a small size to form an event horizon, because it has higher mass. It would need to be compressed to below 3km radius, which is about half the size of the Earth. So the sun would form an event horizon, or black hole, of this size. If the sun were to be compressed in size to 1cm radius, the black hole it produced would still be at 3km, as this is where the gravity strength would be sufficient to attract light.

Scientific studies and simulations have shown that large stars (bigger than our sun) could indeed collapse to form black holes at the end of their lives. We have observed many black holes in the universe of various sizes from very tiny to very large. We can detect them by their strong gravitational effect on stars that we can see, and also by observing 'jets' of matter that line up with their poles due to the strong magnetic fields that they create.

Some physicists believe that huge stars can shrink in size to a point around the same size as this full stop. Recall the discussion earlier about the Earth on a rubber sheet illustrating that the Earth's mass causes space to bend. Now consider that the gravity gradient around a point-sized star would be almost infinitely greater; the indentation in the sheet would be almost infinitely deep as shown in figure 13.

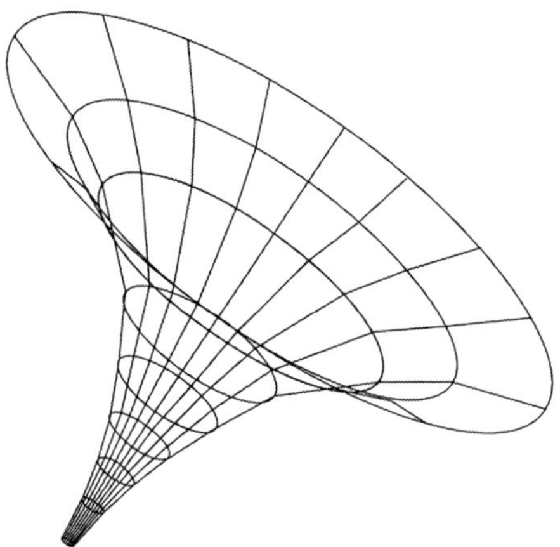

Figure 13. A point-size star causes an almost infinite distortion in space[16]

Figure 13 shows the very deep hole coming in from one direction, but in fact it comes in from every direction and, just like the case with the Earth, it is impossible to imagine this shape in our minds

[16] Graphic courtesy of www.seasky.org

in three dimensions. This kind of distortion or curvature in space is present wherever a point-sized star is present. In theory, these kinds of deep curvatures lead to the idea that one could take a 'short-cut' through such a region of space to get to a far-distant part of the universe. Science fiction has built on this idea, where 'wormholes' can be created through regions of highly curved space that are close to black holes, to take us to far-away regions of the universe very quickly.

This is currently pure fiction; even if we could reach a black hole, the stresses put onto a space-ship attempting such a manoeuvre would destroy it, given the current construction materials that we have. There would be a high probability of being sucked into the black hole; if light cannot get out then neither could anything else. Even if the spaceship survived, its presence would de-stabilise the curved space and the behaviour of any wormholes would become unpredictable.

Coming back to the serious discussion, we have seen that gravity can only be described if we use higher dimensions than our usual set of three. This led to a discussion of black holes that are almost infinitely dense, and generate almost infinite gravitational strength. Now we need to take one small additional step to see how this connects with God. Imagine a clock that gets close to a black hole; it is certain to slow down or stop altogether in that environment. This would happen whatever technology the clock uses; mechanical escapements would stop, electronic oscillators would stop, atomic clocks would stop. Time itself slows and stops. This can also be proven mathematically; time slows down and then stops running altogether for anyone that approaches a black hole. We have seen that space is warped by such objects; the same is true of time.

Now, if we recall our earlier conclusion that God exists outside of our dimensions and outside of our time, we realise that these are exactly the conditions inside black holes. This leads us to the thought that black holes are objects that exist on the very boundary of creation and Heaven.

Matter and particles

We looked at gravity in the last section and saw that it is a force that operates over very large distances, across the universe in fact. Within the scientific community there is also the concept of different kinds of forces that act over very tiny distances. One of these forces relevant to us is the 'strong force', which operates over a distance that is much less than the size of an atom; in fact over the size of an atomic nucleus.

To see this we take the example of the oxygen atom, the structure of which is shown in figure 14. The large central area is the nucleus; it has sixteen spots, eight dark ones and eight lighter ones. The figure also shows two concentric shells outside the nucleus, an inner one with two spots on it, and an outer one with six spots. The spots inside the nucleus represent protons and neutrons, and the spots on the shells represent electrons that are travelling round and round the nucleus. The electrons are orbiting the nucleus in a similar fashion to the planets orbiting the sun in our solar system, although in the case of an atom the attraction is electrical rather than gravitational. Oxygen has a total of eight electrons divided between the two shells. Atoms always have the same number of protons in the nucleus as they have electrons orbiting the nucleus.

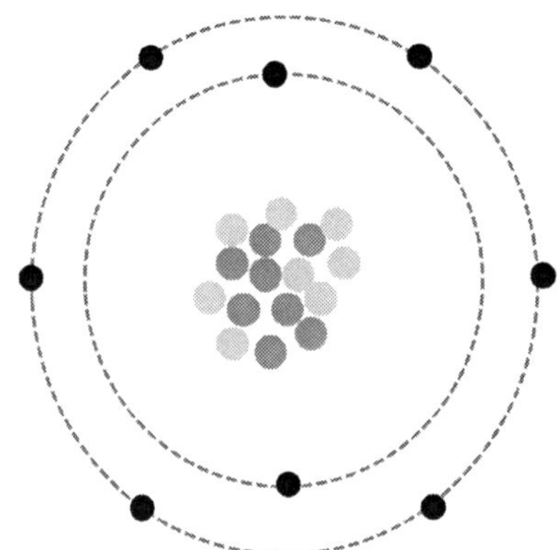

Figure 14. An oxygen atom has eight electrons on two rings around the nucleus

Inside an atom, the electrons always possess a negative charge, and the protons inside the nucleus possess a positive charge of exactly the same magnitude as the negative charge on the electrons. An atom has the same number of protons as it has electrons, in order to achieve zero overall electrical charge across the whole atom.

Inside the nucleus we have the protons and also another type of particle called neutrons, represented by the lighter coloured spots in figure 14. The neutron particles add mass to the nucleus but they have zero electrical charge (hence their name), so that their presence doesn't mess up the overall zero charge in the complete atom. We take the opportunity to introduce them here because they play a part in radio-active dating methods that we discuss later in the book.

Inside the nucleus we have a problem. The size of the nucleus is tiny, about 10^{-14}m across, or a 1 with 14 zeros in front of it (0.00000000000001m). Crammed into this tiny size we have many protons, eight in the case of oxygen. How do they stay so close together? They are all positive charges, and we know from elementary physics that like charges repel each other, much the same as like magnetic poles repel each other. If you bring two magnets together, with their north poles close together, you feel them repelling each other. It's the same with electrical charges. So, how do the protons stay so close together in an atomic nucleus? Well, the truth is, we don't really know, so scientists have invented a force called the 'strong force' that acts over short distances, overcoming the repelling

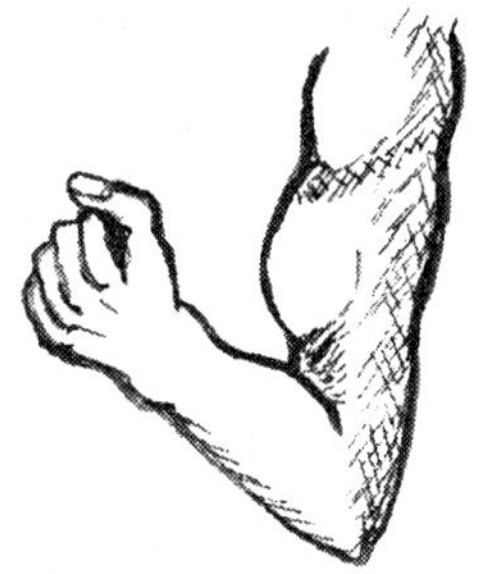

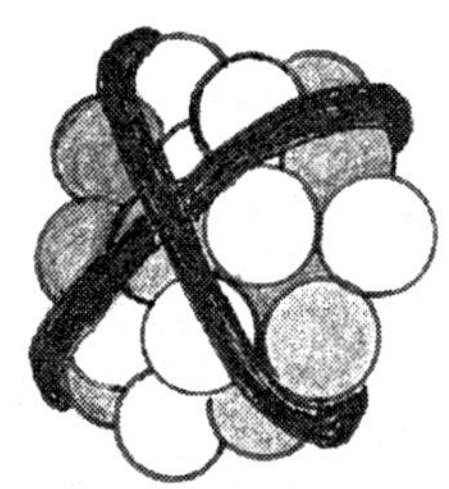

Figure 15. The strong force binds the atomic nucleus together, overcoming the repelling force between protons

force of the like charges. Figure 15 shows the idea, with lighter and darker balls representing protons and neutrons.

The strong force is another phenomenon that cannot be explained by science if we limit ourselves to three dimensions[17]. So atomic nuclei and hence matter itself is held together by a force coming from outside of our three dimensions. We venture to suggest that this force may be considered as spiritual, since science cannot really explain how it comes about. This has support from scripture; the Bible tells us in Colossians 1:17 that '*in Him (Jesus) all things hold together.*'

We therefore assert that Jesus is holding all matter together. We can speculate that, at any point in time, He could release all the atomic nuclei to fly apart and all matter would immediately cease to exist.

Let us look at another aspect of sub-atomic particles that proves they operate in higher dimensions. For this we choose a different particle as our example, the electron. We have come across electrons already, orbiting the nuclei within atoms. It is straightforward to produce electrons on their own, outside of atoms, so that studying them in isolation is fairly easy.

Electrons behave in unusual ways. One example of unusual behaviour is that, by using specialised laboratory equipment, you can turn an electron all the way round, full circle, and it looks different to how it did when you started. You have to turn it all the way round a second time before it looks the same again. It is known as a spin-half particle because you have to turn it all the way around twice before it looks the same again. The electron is not the only particle that displays this very strange behaviour; there are many others with this kind of 'fractional' spin.

Another example of the unusual behaviour of the electron can be displayed by making use of a piece of laboratory equipment called a

---

[17] The strong force is theorised to come from quarks that make up the protons and neutrons and overflows to adjacent protons and neutrons Quarks may themselves be made from even more fundamental particles called strings.

diffraction grating, which is a kind of screen with very closely spaced slits in it. Electrons, even when sent through the grating one at a time, can be observed to pass through using multiple paths at the same time. Have a look at figure 16.

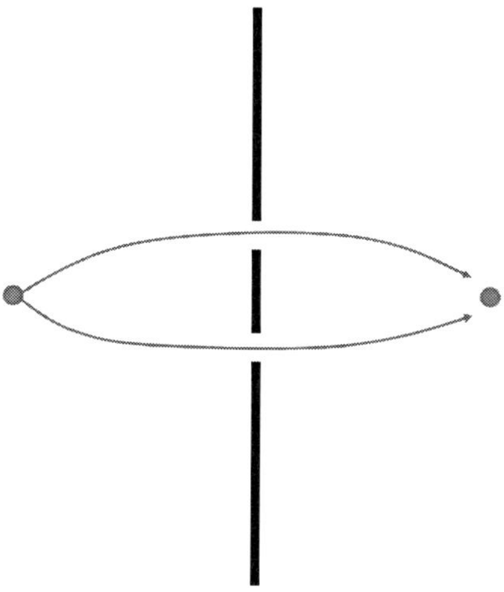

Figure 16. An electron sent from one side of a diffraction grating to the other will go through every available path

If the grating has many slits, then the majority of the energy of the electron goes through the most direct routes, and gradually less energy goes through the less direct routes.

Now let us consider if this idea can be applied to much larger objects, such as human bodies. If you or I walk from one room into another, we would normally go through a door, and anyone can observe us going through the door. But in reality, if we move between any two locations, just like the electron, a tiny bit of us travels on every possible path between the two locations. Every possible path has what is called a 'probability' associated with it. By far the highest probability path is through the door, so that is the way the vast majority of our body went. In truth, it is a bit blurred around the edges, and a very tiny amount of our body travels by every other possible route. There is no need to worry about this; in the case of our bodies the amount of material that travels on side-routes is so small that it can safely be ignored.

However, as objects get smaller the probability distribution is not so sharp, and the blurriness increases to the point where very small objects can be observed to travel using multiple paths very clearly, as we have seen with the electron. So be ever mindful, as before, that what you observe or experience is never the whole story.

We can only wonder at how our bodies, and indeed at all matter, are tied up with God. We know that matter is composed of atoms, and every atom is composed of a nucleus and electrons. This means that all matter contains a very large number of particles that need higher dimensions than the familiar three to describe them. Whether we realise it or not, all matter, including our bodies, are intimately entangled with God.

How many dimensions do these tiny particles such as electrons actually have? A theory that is gaining much traction is 'string' theory, which basically says that all types of particles consist of vibrating tiny strings, and that the difference between the particles (electrons, protons, neutrons, and others) is due to different vibration patterns. The strings vibrate in many different dimensions, not just in the three dimensions that we are familiar with; the popular thinking is that they operate in eleven dimensions. We don't know yet whether string theory is right, but it does, so far, fit all the observations on the behaviour of matter and forces.

There is even some support for string theory in the Bible, where reference is made to '*morning stars singing together*' in Job 38:7, where we can imagine the vibrations and harmonies in a huge orchestra with the Creator as the conductor.

Summarising this section, we have seen that gravity, the forces that hold matter together, and sub-atomic particles like electrons, can only be described by a set of dimensions that is higher than the familiar three of length, width and height. We have seen that black holes have the same conditions internally that exist in Heaven. All matter seems to be intimately connected to Heaven and is probably a shadow of it. All of this is backed up by scientific observations and by evidence from the Bible. We assert that God is holding everything together.

Finally in this section, let us return to the question of what is meant by the term 'universe' and modify it slightly in the light of the discussion. Let us now agree that the universe is all matter and life associated with it, regardless of how many dimensions the matter actually has.

Point to ponder: development of string theory has been very slow and difficult mathematically but nevertheless some progress has been made in determining what kinds of vibrations are responsible for certain characteristics of particles. The string theory around some of the larger particles, such as electrons, has been proved using particle accelerators. The particle that is supposed to be responsible for gravitational force, the graviton, exists in string theory, but to prove it exists in practice would require a far larger particle accelerator than is currently available.

The Earth is in a special place

Now let us discuss briefly the relative position of the Earth and the stars. In history of mankind we have been through a phase of believing that the Earth is in a special place in the universe, even at the centre, and then into times of believing that the Earth is not in a special place. The Catholic church believed the Earth was at the centre of the solar system, and indeed at the centre of the universe, until proved wrong by Copernicus and Kepler in about 1620. Since then, astronomers have shown that the Earth is not in any particularly special place. However, we will try to show that the Earth is in a special place after all, and yet not in the centre of anything.

When we look at the stars on a dark night, it seems as though there are more stars in some directions than in others, but this is because we can see only a few of the closest ones with our naked eyes. If we look with more sensitive instruments, and with instruments that detect not only visible light but also infra-red and radio signals from stars, we discover that the number of stars is roughly the same in whatever the direction we look. In this sense it might seem that we are in the centre, but in reality there is no centre; we would observe the same thing wherever we were in the universe. Recalling the illustration of an inflating balloon, we would be somewhere on its surface and there is no 'centre' of the universe to be located at. So, how are we to justify our belief that the Earth is in a special place if there is no centre?

We begin the discussion by considering the speed of light. Light travels very quickly, but not infinitely fast; we can measure its speed in the laboratory at just under 300,000km per second. This means that it takes roughly one-tenth of a second for light to travel right around the Earth's surface, and roughly ten minutes to travel from the sun to the Earth.

Now consider a distant star so far away that its light takes several years to reach us. We measure the distance to such stars as so many 'light-years' away. In fact the most distant stars that we can see with radio telescopes are several billion light-years away, a truly enormous distance. This means that the light we are seeing from these stars started out several billion years ago. Using these numbers, astronomers can estimate the age of the universe from the stars that are furthest from us, and currently the estimate is 20 billion years.

How can this estimate be reconciled to the Bible account that creation happened only six thousand years ago? This is the famous 'starlight' problem that is often presented as a challenge to believers in the Genesis account of creation. Now we believe that the Creator can start the universe off in any way He wishes, including the creation of starlight already in transit if need be. Also, we have seen that God is stretching out the heavens, and it is possible that the rate of stretching in the past was much greater than it is now. But let us hold these thoughts while we recall that we are also grappling with the thought that the Earth is in a special place.

Now also bring into the mix the scripture from Genesis 1:16 that says *'He also made the stars'*. The Hebrew for this scripture could also be translated 'the stars became visible', which for our argument here is the better translation. For this to happen, all stars becoming visible at the same time, there is only one place that the Earth could be, because of the vast differences in the positions of the stars. The stars becoming visible, all together at the same instant, could only have been observed on the Earth if it was at a special position at that instant in time. Whether the universe was stretched out very quickly at the beginning or whether the starlight was created in transit, we must have been in a special place to have seen all the stars appear together. An observer who happened to be on Mars instead of the Earth, for example, would have seen some difference in time between the first stars appearing and the last ones appearing. But the Bible says that on the Earth they all appeared together. God surely must have orchestrated this, because the simultaneity of the stars appearing depended upon the Earth even being in a particular orbital position.

For those who struggle to believe that the stars appear to be older than they actually are, note that God has done the same thing on other occasions. He is good at making things appear older than they actually are. When Adam and Eve were formed, they were formed as adults. On another occasion, Jesus turned water into wine at a wedding in Cana, which was so good that the wine-waiter commented upon it, according to John 2:10. Whereas a young wine can be good, it is surely necessary here that wine only 30 seconds old tasted a lot older!

The big bang has big problems

One scientific theory that was popular at one time, the big bang theory, has big problems both from the scientific and biblical viewpoints. This theory runs as follows. An infinitely hot and tiny ball of radiation, much like a huge black hole, exploded and expanded into the universe that we now see. This infinitely small ball contained all the mass and energy of the universe and also space itself. The tiny ball expanded, spreading out the mass, energy and space, much like our balloon in figure 8 expanding from zero size. Some of it turned into matter, mainly hydrogen and helium gas, which collapsed to form stars. The energy, in the form of heat and other radiation, would have been too high for heavier elements to form. As the stars got old and ran out of hydrogen to burn, they created heavier elements in their cores. At the end of their lives the stars exploded, sending these heavier elements into space. New generations of stars would continue to be born from clouds of hydrogen. Some of the heavier elements began to stick together and whirl around the new stars, forming the Earth and the other planets. Let's see briefly why this theory is wrong from the scientific viewpoint, and then from the biblical viewpoint.

First, particle physicists tell us that many magnetic monopoles would have been formed in such high temperature conditions. These are massive particles and they have only one magnetic pole, unlike the more familiar two-pole magnets that we have on our refrigerators. These monopoles are stable and should have lasted to this day, but they have not been found despite considerable search efforts.

Second, the balance between the expansion of the universe and the gravitational attraction across it is a very fine one; this condition is called 'flatness'. A high degree of flatness is needed to prevent the universe from rapidly collapsing in upon itself or from rapidly flying apart. Flatness decreases with time, as this is a type of order, and we know from experience and from the second law of thermodynamics that the order in any system decreases with time. Therefore in the past the degree of flatness must have been even finer than it is today, and for this to happen all the way back to the big bang it must have been flat to an extremely high precision, despite the fact that the laws of physics allow for an infinite range of values. This is a co-incidence that is beyond credibility.

Third, there is absence of anti-matter. Any creation of energy into matter would have generated equal amounts of matter and anti-matter. In reality the universe is comprised almost entirely of matter, with only trace amounts of anti-matter having been detected.

Fourth, the big bang would produce the only the lightest three elements, which are hydrogen, helium, and lithium, because the conditions would have been too hot for any larger molecules to hold together. The theory says that the early stars would produce the first heavier elements only after running out of hydrogen fuel. Some stars, the very early ones that have not yet run out of hydrogen, should consist only of only these three lightest elements. However, all stars, even the very early ones that have been around from the very beginning, have heavier elements in their cores.

The biblical problems for the big bang are many but we will only look at four to keep the number the same as the scientific problems. First, the account in Genesis tells us that the Earth was formed before the sun and the stars, not afterwards. Second, the account in Genesis tells us that creation happened in six literal days, and this allows nowhere near enough time for the universe to expand from a tiny ball and settle down enough for human habitation on the Earth. Third, the account in Genesis teaches that the Earth was made under water, whereas the big bang says that it started out as a molten blob of heavy elements from an exploded star. Fourth, the big bang theory predicts that the universe will either carry on expanding indefinitely (the open universe model), eventually stop expanding (the flat universe model), or start contracting in upon itself (the closed universe model), in a future timescale that is billions of years. The Bible account of the end is very different, the Earth and the heavens will roll up '*like a scroll*' as described in Revelation, in a timescale that is comparatively very short, just over a thousand years from now.

More on light and time

Light and time are subjects that have been studied scientifically very deeply, and there is some scary mathematics that is used to describe their behaviour. We won't go into this kind of depth, but we would like to point out a few more of the characteristics of light and time that will make it clear they were designed by God as types of boundaries for our physical bodies.

After the creation of heavens and the Earth, light and time were the two next things that were created:

> *'Then God said, 'Let there be light'; and there was light. And God saw the light, that it was good; and God divided the light from the darkness. God called the light Day, and the darkness He called Night. So the evening and the morning were the first day.'*
>
> GENESIS 1:3-5

So God created light, and then He started time flowing or the clock ticking by causing evening and morning cycles to start up.

We have said that light travels at just under 300,000km per second. Now if we try to travel that quickly, we begin to encounter some interesting effects. If we accelerate any object towards the speed of light, something curious happens: the object gets gradually heavier. As it gets heavier it requires more energy to make it go faster still, which makes it even heavier. As it approached the speed of light the object would get infinitely heavy, and so it would need infinite energy to actually achieve the speed of light. In this way, the speed of light is a kind of barrier that we cannot overcome physically. This is illustrated in figure 17.

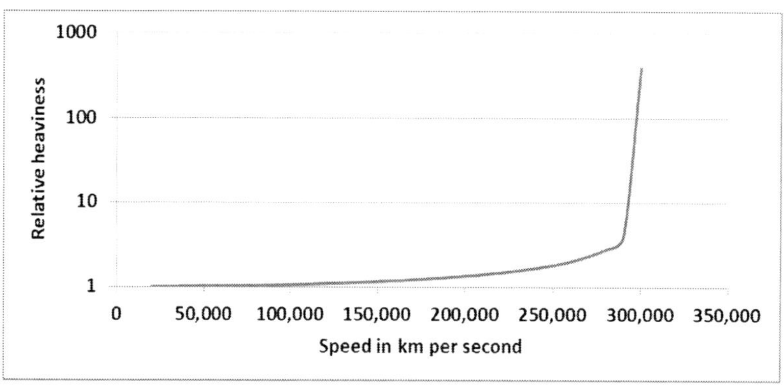

Figure 17. The heaviness of an object increases as its speed approaches the speed of light

Figure 17 shows that the heaviness of an object would increase by one percent at about 50,000km per second, and at a speed of 260,000km per second the heaviness would have doubled. This means that the energy needed to make the object go faster still would also be doubled. The heaviness would double again when the speed reaches 290,000km per second. As the object approaches the speed of light at 300,000km per second its heaviness would increase steeply to infinity. This means that infinite energy would be needed to accelerate the object past the speed of light. As it is impossible to produce an engine to give infinite energy, it means that nothing physical can ever exceed the speed of light. This is true for all objects, and it is also true for any kind of signals that carry information. So the speed of radio or television or cellular telephone signals, which carry information, cannot exceed the speed of light, indeed these signals travel exactly at the speed of light. Effectively, it puts a limit to how far people can travel in a lifetime. It is a barrier that keeps us constrained.

What about time, does that also behave strangely as we go faster and faster? Yes it does. It is intimately coupled with light. To show this, have a look at figure 18.

Figure 18. The image departs from the clock face at the speed of light

If we were to travel away from a clock at the speed of light, we would carry with us the image of the clock face, and we would not see further changes in time that the clock is indicating. This means that for us the clock would not move and time would stand still. You may be sceptical about this, but experiments and calculations have shown this to be the case. Time would stop for us if we could travel at the speed of light.

This effect of time slowing down comes on gradually with increasing speed, and it is not necessary to travel very fast for the effect to be noticed. Scientists have flown atomic clocks in aircraft, brought them back to the ground, and compared them with clocks that were on the ground the whole time. The ones in the aircraft were showing less time had passed than the ones that stayed on the ground; the moving clocks had run slower. It is the same for humans. If we travel at high speed, our body clocks slow down and we age at a slower rate than someone who stays at home. Of course the effect is very slight the kind of speeds we travel at, so don't get excited at the prospect of staying young by living on an aircraft.

So now we have seen two ways to stop time in this chapter. First we can be in a very strong gravity field close to a black hole and, second, we can travel very fast, approaching the speed of light.

While living here upon the Earth, we are constrained by these barriers of light and time. We are convinced that this light-time system is designed to constrain us here on Earth, or very close to it, while we are alive. God has set this up as a kind of reference structure, and we believe that its purpose is to keep us safe and sane.

Now we are in a position to clarify another relationship between the physical and spiritual realms. Our spirits have zero mass, so that when we die and our spirits leave our bodies, we can pass through the light and time barriers. Our spirits will therefore have the ability to travel infinitely fast, so they can go anywhere including into Heaven or Hell. Time will also stop, which means that our spirits will be around for eternity. When this happens, one day will feel the same as a thousand years, just as the Bible tells us in 2 Peter 3:8.

Things get even stranger when we look even deeper at light and time, especially when we examine their interaction with gravity and curved space that we discussed earlier. Light is affected, or 'bent' by curved space that is caused by massive objects, so that all the stars we see in the distance are not quite in the positions that they appear to be. The concept of a 'straight line' does not exist in reality, since space itself is always curved, if only slightly. Light follows the shortest path between any two points, which is always a curve. Yet again we come across the concept that things are not quite as they seem, and there is always more detail to consider. How vast and yet how subtly detailed is God and His creation.

Point to ponder: Imagine you are travelling at ninety percent of the speed of light. Coming towards you in the opposite direction is someone else who is also travelling at ninety percent of the speed of light. Logic leads you to the conclusion that you are approaching each other at nearly twice the speed of light. But in fact both of you will measure the speed of the other person at less than the speed of light relative to your own speed. This is because your clock is running slower than it would be at rest, and also whatever you use to measure distance would be longer. This is what Albert Einstein modelled in his special theory of relativity: nothing can travel faster than the speed of light, regardless of the speed of any observer.

The Flood and Other Events

The Earth is not a stable environment. It has been subjected to a major flood and other cataclysmic events in the past. Here we present some evidence for a few of them.

A worldwide flood

Here we present evidence that there has been a worldwide flood or deluge that significantly changed the atmosphere and radically re-arranged the surface features of the Earth in the not too distant past. The Bible tells the story of the flood in Genesis chapters 6-9, where Noah and his family were the sole human survivors. Noah built an ark or boat, and inside this ark lived eight people during the flood. These were Noah and his wife, their three sons and their wives. Also aboard were seven pairs of every 'clean' creature and two pairs of every 'unclean' creature. During the flood, all land-based creatures including humans were destroyed; only those aboard the ark were saved. Importantly, all the giants on the Earth also perished.

There are many questions related to the story of such a flood, which we will attempt to address. Some questions that come to mind might be these: Where did all the water come from and where did it go afterwards? How high did the water get? Did it cover the whole Earth all at once? How did all the animals get to Noah's ark and how were they distributed afterwards? The answers may be found by performing a little research into geology and from studying carefully the book of Genesis. Let's start by taking a look at what the Bible says regarding the source of the water:

> *'In the six hundredth year of Noah's life, in the second month, the seventeenth day of the month, on that day all the fountains of the great deep were broken up, and the windows of heaven were opened. And the rain was on the earth forty days and forty nights'*
>
> **Genesis 7:11-12**

This scripture tells us that some, probably the majority, of the water came from underneath. Scripture tells us that there was much water below the surface of the Earth, which burst forth during the flood. This would have been very hot water, heated by the close proximity of the Earth's mantle that is at a temperature of several hundred degrees centigrade. This bursting forth was very violent and, during this process, some water and rocks may well have achieved escape velocity and been flung off the Earth into space. This means the rocks that now make up the comets and perhaps even the asteroid belt could have originated on the Earth. Scientists have recently discovered ice on the poles of the moon; they are struggling to understand out how water got there. Well, we can surmise that it arrived there during the flood.

According to the creation account, the Earth had water above the firmament (sky) as well as on the surface:

> *'Then God said, 'Let there be a firmament in the midst of the waters, and let it divide the waters from the waters.' Thus God made the firmament, and divided the waters which were under the firmament from the waters which were above the firmament; and it was so.'*
>
> GENESIS 1: 6-7

There is a scientific debate on the Internet about whether the Earth could have had a water 'canopy' that is implied by the scripture above. Scientifically it is possible; at an altitude of 50,000 feet water boils because the pressure of water vapour is the same as atmospheric pressure at that height. The vapour would be transparent because it would not condense in the way that steam condenses when coming from a kettle.

A beneficial aspect of such a canopy is that it would have afforded greater protection from the harmful rays from the sun, which could be one reason for people living longer before the flood. Such a canopy does have some issues though. First, its presence would increase the atmospheric pressure and the trapping of heat. Second, when the canopy came down during the flood, the potential energy possessed by the water (through it being high up) would be turned into heat. This would have raised the temperature at the Earth's surface by an amount depending upon the thickness of the canopy, and the rate at which it fell.

Due to changes in the atmosphere and the environment, the life span of humans has been shortened considerably. Noah built the ark when he was over 500 years old, and he lived to over 900 years old. Others before him lived a similar length of time as recorded in the book of Genesis. Today the lifetime of humans is a maximum of 120 years and typically much less.

Our thinking on the canopy is that one did exist, according to the Genesis account, but it was thin enough not to cause excessive pressure or heat problems. Bearing in mind that most of the water came from underneath the Earth, the canopy did not have to supply much of the floodwater and could have been just a few centimetres thick. Also, the Genesis account tells us that rain fell for forty days, so the canopy came down slowly and may not have caused significant heat problems as it descended.

It is clear that the atmosphere changed very significantly during the flood, because before the flood there were no rainbows. Indeed there was probably no rain as we understand it; we are told in Genesis 2:6 that the ground was watered by a mist that rose from the land. Recall from chapter 5 that the rainbow is the token of the covenant between God and Noah and all those on the ark with him. The removal of a thin canopy could explain these changes.

The water coming from underneath was hot and the water falling from the canopy was probably also hot by the time it reached the Earth's surface. Significant loss of life would have occurred in the sea and rivers due to the ingress of an enormous amount of hot water rich in minerals. On the land, the rapid burying of forests and creatures by mineral-rich hot water would have resulted in the fast formation of coal and fossils as we shall see later.

When the water was accumulating for the flood, the Earth may have changed size slightly and the surface fractured, producing water from the '*fountains of the deep*' and at the same time the canopy would have come down in the form of rain. If the water spurted with very high pressure from the ground, high enough for some to spurt into space, then large amounts would have fallen back as more rain. After the flood the Earth re-stabilised, changing size, making enough room for the water to settle into the oceans that we now have. The disturbance in the Earth structure may explain why the orbit period (the year) slowed down from 360 days in early Old Testament times, to the current period of 365.25 days.

Bear in mind that before the flood all the seawater was in one place according to Genesis 1:9. It isn't all in one place now, so clearly there has been a substantial re-arrangement. As all the sea was in one place, then so was the land. All the creatures must have been close together on that one piece of land at the time of creation, because God brought them all to Adam to see what he would call them:

> *'Out of the ground the Lord God formed every beast of the field and every bird of the air, and brought them to Adam to see what he would call them. And whatever Adam called each living creature, that was its name. So Adam gave names to all cattle, to the birds of the air, and to every beast of the field.'*
>
> **Genesis 2: 19-20**

The creatures taken aboard the ark were all on this one piece of land, so no crossing of seas was needed. When the flood was in progress and also afterwards, the land mass split up and different chunks travelled across the surface of the Earth. This gives us one possible mechanism that caused animals and people to become distributed. Mountain forming took place as a result of the re-arrangement of the Earth's surface, to get to the situation we now observe of oceans and land masses, including mountains, rising above them.

The bursting of the Earth's surface flung rocks and water into space, the reactions were so violent, and we have said this could be the source of the comets. Halley's comet, for example, has an orbital period of about seventy years and its orbit is very elliptical. It is made of dust and ice. Every time it approaches the sun it loses some of its mass due to melting of the ice. The thought that this comet has been around for billions of years is not reasonable, losing mass on many millions of orbits. We have also heard the theory that the comets came about through collisions in the asteroid belt. But the more likely explanation is that the comet was born from the Earth during the flood a few thousand years ago.

As for the depth of the waters, we know from the Bible that it covered the highest mountain, but the highest mountain at that time was probably Ararat where the ark settled when the waters were receding. The higher mountains of today simply may not have been there before the flood. There is evidence that the mountains of Nepal including Everest were not there before the flood: like many mountains they have marine fossils and shells even on their summits.

We are convinced that the flood was not a gentle affair of water appearing and then draining away, but a very violent cataclysm, changing the structure and the size of the Earth, sending material into space, changing the atmosphere and re-arranging the surface features.

At first you may think that this is beyond credibility. Yet it starts to make sense when you start to consider some evidence for it. When you start looking for the evidence for the flood it becomes convincing, and you wonder why you didn't notice it before. Let us now look at some.

Rocks

Sedimentary rock is formed in layers, where water or mud carries particles of sand or other materials and lays them on top of other layers, and the height builds up. By its nature it forms horizontal layers when formed. Sandstone and limestone are common examples of sedimentary rock. Take a look at figure 19. The picture, taken in Israel, is of sedimentary rock that has been buckled so badly that some of the layers have become vertical. There are examples of this kind of buckling all over the surface of the Earth.

Figure 19. Sedimentary rock is buckled like this all over the Earth

We are all told at school that building of sedimentary rock was a process that took millions of years, where gentle tides build up the silt and mud. We are told that buckling of the rocks was caused by movement of the Earth's tectonic plates and up-thrusting geological activity, over a period of millions of years. One difficulty with this theory is that the rocks would surely shatter if subjected to such pressures and movement.

Now let us suggest an alternative theory. Consider a world-wide flood that swept countless millions of gallons of hot water and silt and mud across the Earth, burying everything, building the sediment quickly in a matter of weeks. Then came a buckling under the enormous pressure from the massive re-distribution of the Earth's surface as the water subsided, which folded the layers while they were still wet, hot, and flexible. We encourage you to stare at such formations when you next come across them and see if you don't agree that this is the more likely explanation.

We also observe that, all over the Earth's surface, there are large rocks that are out of place. These can be observed everywhere if you look out for them; rocks that have been carried many hundreds of miles and dumped onto other rocks of a different type. These out of place rocks are called erratic boulders or simply 'erratics' and some of them are huge indeed.

Figure 20 shows examples in Canada and Estonia. They can be found on every continent; in England there are some in Norfolk and on Dartmoor.

Figure 20. Examples of erratics. Left is Big Rock in Canada and right is a granite boulder called Ehalkivi in Estonia[18]

Something carried these rocks long distances and dumped them where they are now. One theory we have heard is that they were carried by glaciation, but to us it seems more likely that the agency responsible for transporting them was not ice but water, a great deal of water.

Have a look at figure 21. It is a lateral view of the stratified volcanic rocks in the United States around the Grand Canyon. On the scale of the picture, the Colorado River is less than one pixel in cross-section. It is tiny, and the idea that it cut out the canyon is beyond feasibility. The river flows in the canyon, but it did not form the canyon; the river began flowing in recent times. We know this because of two main characteristics but there are others. First, if the river had cut out the canyon, the mud and debris would have ended up as a river delta. The Colorado has no delta, and there is no evidence of it depositing the mud and debris anywhere. Second,

[18] Image credit: Dolce Vita and Dainis Derics / Shutterstock

the top surface of canyon itself goes uphill across the Kaibab plateau in the direction of the flow and so the Colorado would have had to flow uphill to carve it out. So what did form the Grand Canyon? Was it perhaps a violent re-arrangement of the Earth's surface, where the landscape was fractured and uplifted due to stresses caused by ingress and egress of large volume of water?

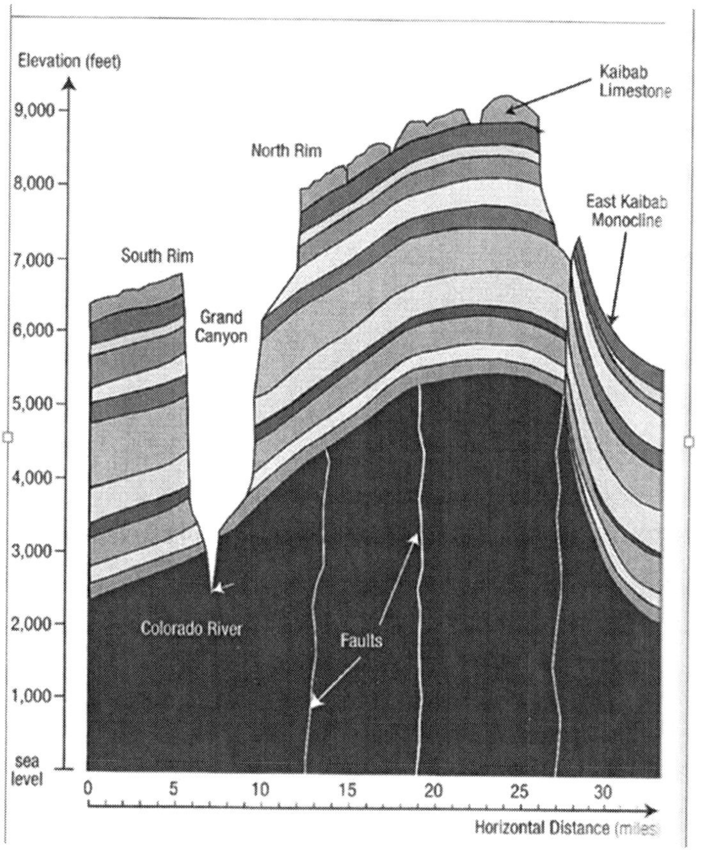

Figure 21. The Colorado River could not possibly have cut out the Grand Canyon[19]

[19] Graphic acknowledged from *In the Beginning: Compelling Evidence for Creation and the Flood*, 8th edition, 2008, Walt Brown.

Fossils are formed quickly

Mount St Helens in the United States erupted in 1980. The eruption lasted several days and buried the trees around it. Several hundred feet of rock were laid in a few days, stratified just like the Grand Canyon. The trees were fossilised within a few weeks, before they had time to decay. This eruption has demonstrated that stratification of rocks and formation of fossils are processes that can happen over a timescale of weeks, and do not need millions of years.

Figure 22 shows a tree that is fossilised in several layers of rock. This could only have happened if the rock formed around the tree, and did so very quickly, because the delicate features of the tree remain visible. You can even make out the flow lines as the silt made its way around the tree. These trees are called 'polystrate' because their fossils extend through several strata of rock. Many of these trees have been discovered; some are even upside-down because they were uprooted by the violent water and mud flow. This observation does not support the 'accepted' view that says fossils are formed over timescales of millions of years.

Figure 22. A polystrate tree. It is several metres tall and is fossilised through several layers of rock

This example and the Mount St Helens example show that the fossilisation process can happen very quickly if the conditions are suitable. Fast burial in hot mud and water with high mineral content would have been such suitable conditions brought about by the flood.

There are many examples of animal fossils that indicate fast burial and fossilisation. Two such examples are shown in figure 23.

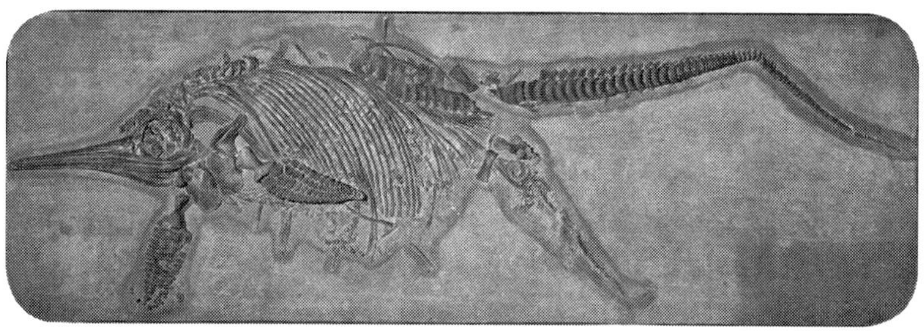

Figure 23. Fossils of an Ichthyosaur giving birth to a baby and a fish eating another fish[20] [21]

The amazing fossil of the Ichthyosaur shows a mother giving birth. One explanation of how this fossil came about is that both mother and baby died while giving birth. Whereas it is remotely possible that they both died naturally at the same time, a far more likely explanation is that they were buried rapidly at the point of birth. The fish eating fish in figure 23 indicates rapid burial of the fish by water and mud, before the larger one finished his meal. There are many more such images on the Internet. Evidence of high speed high pressure burials is common among fossils, and the inference is

[20] Ichthyosaur image © Natural History Museum, London

[21] Fish image used by permission of Answers in Genesis, www.answersingenesis.org

that they were buried by a great deluge of mud and water, such as from the flood, rather than by a slow process of silting over.

Further indications of fast and violent burials are obtained from fossilised mammals. Fossils of hundreds of mammoths have been found in the United States and Siberia that still have soft tissue such as hair, and have food in their stomachs. This means that they did not have time to decay before they were fossilised. We encourage you to look these up on the Internet.

Many fossils are found with the head arched backwards and the tail bent around almost on top of the head, as shown in figure 24. It seems that they were they caught suddenly by a rush of water and mud, and were drowned and buried quickly.

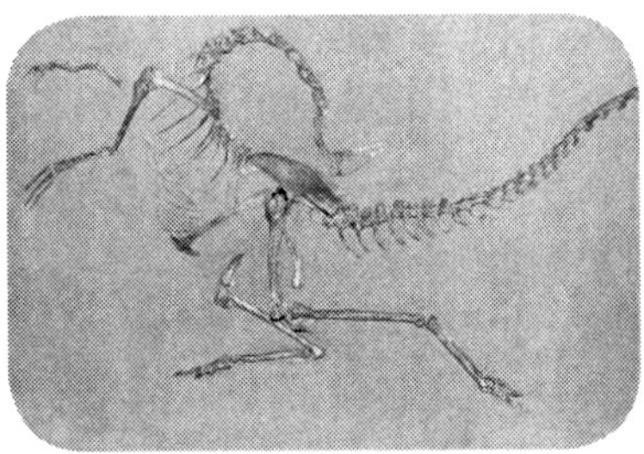

Figure 24. Many fossils found in this kind of position. Caused by fear, running or sudden drowning perhaps[22]

Here is a quotation from an evolutionary standard text that was written to explain a discovery of bones in a cave in Settle, England:

> *'Victorian excavators were particularly fascinated by 'bone caves' where there might be a possibility of finding evidence for the earliest humans along with long extinct animals. Victoria Cave certainly had plenty of animal bones. The earliest, at 130,000 years old, included those of hippos, narrow-nosed rhino, elephants and spotted hyenas. They date to an Upper*

[22] Images credit BG Smith and IPK Photography/Shutterstock

> *Pleistocene interglacial when the climate was much warmer than today. It seems as if at that time, the hyenas were using the cave as a den and dragging scavenged bones back to it. No evidence was found for human activity during this period.'*[23]

This account has several suspect and unreasonable ideas, such as Hyenas dragging complete remains of animals much bigger than themselves into the cave, and the dating of bones using Carbon 14 with its unreliable assumptions that we discuss later. The account also assumes that these animals lived in England at one time; this is a remote possibility, but a more likely explanation is that they were carried a long way and washed up there. The text goes on:

> *'The glaciers then returned and from 120,000 to around 12,000 years ago the cave gradually filled with layer upon layer of clay deposited as the glaciers periodically melted.'*

Our alternative flood-related theory is that the clay would have arrived very quickly, brought by wave after wave of hot water, rather than being deposited by repeatedly melting glaciers. Besides the cave in Settle, bones have been found like this in other caves in England and Wales, and the story is repeated all over the world. Caves contain bones from animals not usually found together, some predators of others, roughly all the same size.

More generally the fossils we find are of animals that died and were buried quickly, mostly in positions of agony or fear. Some are in herds that are found all facing the same way in running postures. Fossils are found at various depths often sorted by size, and never by 'evolutionary' order.

This is only the beginning of the evidence; it is impossible to list all the evidence for the flood in this book. If you are curious enough

[23] See for example on the Internet at www.outofoblivion.org.uk/record.asp?id=506

to seek out more evidence from the Internet and whenever you travel, you will find overwhelming amounts of it. We are convinced that you will be convinced.

Dinosaurs and large insects

We believe that pairs of every living species were taken aboard the ark except for the water-dwellers. The higher atmospheric pressure before the flood may have been necessary to sustain the respiration of some creatures that have died out since. We have ample fossil records to show that land-based dinosaurs and giant insects were around before the flood. Examples are dragonflies 'meganeuridae' with a wingspan of 45cm and of course the well-known larger land-living dinosaurs.

Figure 25 shows a fossil of a giant dragonfly with a wingspan of about 60cm, along with an artist's impression of what it looked like.

Figure 25. Fossil of pre-flood giant dragonfly and artist's impression[24]

[24] Image credit Graham Cripps, Natural History Museum, London

We are told that dinosaurs have been extinct for millions of years, but we challenge both their extinction and the timescale. There have been reports of dinosaur sightings in Africa in the past few years. Mokele mbembe is the name given by natives of the Congo to a sauropod type dinosaur with a small head, a long neck, and an elephant type body. As for the timescale, there is fossil evidence that mankind and dinosaurs were on the Earth at the same time, which can easily be found on the Internet. It is interesting that the Bible talks of giant creatures called Behemoth (in Job 40:15 – 24) and Leviathon, which can breathe fire:

> *'His sneezings flash forth light, And his eyes are like the eyelids of the morning. Out of his mouth go burning lights; Sparks of fire shoot out. Smoke goes out of his nostrils, As from a boiling pot and burning rushes. His breath kindles coals, And a flame goes out of his mouth'*
>
> **Job 41:18 – 21**

Point to ponder: it was Huxley in 1841 that introduced the term 'dinosaur', before this they were known as dragons. Were there once fire breathing dragons around?

Meteorite impacts

In addition to the changes of the Earth due to the flood, observations show that the Earth has been subject to sudden changes in the environment brought about by external agencies, such as collisions from meteorites. One of the most well-known observations of these is the giant crater in Arizona, shown in figure 26.

Figure 26. Meteor crater near Flagstaff, Arizona, the world's best preserved impact crater[25]

This crater was caused by a small meteorite (less than 100m diameter) composed mostly of iron. The date of the impact is estimated at 50,000 years ago; this estimate is made by assessing the degree of erosion of the limestone rocks that were thrown up and exposed to the open desert weather when the meteorite impacted. We believe this to be an unreliable estimate because it assumes that the rate of erosion has been constant. The crater has layers of sedimentary rock that were uplifted on impact, so it was after the flood that the meteorite hit, less than five thousand years ago.

Another, much larger, meteorite with a high Iridium content struck the Gulf of Mexico that resulted in the Chicxulub crater. This impact would have created a giant ash cloud that spread out around the Earth. The area around the Chicxulub crater is rich in Iridium and this impact may be responsible for the appearance of the so-called K-T (Cretaceous – Tertiary) boundary. This is a thin layer of rock that goes all the way around the world and contains eleven times more Iridium than the rocks above and below it. In textbooks the K-T boundary is usually dated at 63 million years ago, and the

[25] Image credit Action Sports Photography/Shutterstock

Chicxulub meteorite is sometimes blamed for the extinction of the dinosaurs. We dispute the date of the impact, believing it to be a few thousand years instead, but agree that such an impact would have caused climate change for a period of time.

Anomalies in time

In the Old Testament, there are at least two cases where stretched days have been experienced, which can only be explained by a disturbance in the Earth's rotation.

The first is when the sun's shadow moved back ten steps up a stairway in the days of Hezekiah:

> *'And Hezekiah said to Isaiah, 'What is the sign that the Lord will heal me, and that I shall go up to the house of the Lord the third day'? Then Isaiah said, 'This is the sign to you from the Lord, that the Lord will do the thing which He has spoken: shall the shadow go forward ten degrees or go backward ten degrees?' And Hezekiah answered, 'It is an easy thing for the shadow to go down ten degrees; no, but let the shadow go backward ten degrees.' So Isaiah the prophet cried out to the Lord, and He brought the shadow ten degrees backward, by which it had gone down on the sundial of Ahaz.'*
>
> 2 Kings 20: 8 – 11

Isaiah 38:6-8 tells the same story, describing how the sun went back about ten degrees on the sundial of Ahaz as a sign to Hezekiah that he would have victory over the king of Assyria, and that he would have fifteen years added to his life.

The second is where Joshua prayed for more time when he was fighting with the Amorites:

> *'So the sun stood still, And the moon stopped, Till the people had revenge Upon their enemies. Is this not*

> *written in the Book of Jasher? So the sun stood still in the midst of heaven, and did not hasten to go down for about a whole day.'*
>
> **Joshua 10:13**

Joshua 10:11 speaks of the Lord casting giant hailstones upon the Amorites, but the Hebrew word here is the normal one for 'stones'. One interesting theory is that hot stones fell as the tail of a comet swept past the Earth. The mass of a comet, if it had a close encounter with the Earth, would have upset the Earth's rotation and could account for the effects described in the scripture above.

Immanuel Velikovski (who is a scientist and not a Bible scholar) has compiled evidence that such a comet did indeed pass by the Earth at a close distance. In his book 'Worlds in Collision' he forms a theory that the comet was the planet Venus. He claims that its close encounter with the Earth caused many unusual events such as the parting of the Red Sea in the time of Moses, the very long day in the time of Joshua, and portents in the sky such as pillars of cloud and fire. His evidence shows that the sun used to rise in the west and set in the east, indicating that at some stage the Earth turned pole over pole. He also presents some interesting evidence that the atmospheres of Venus and Earth interacted, and hydrocarbons were formed in such great quantities that they produced much of the oil we now have on Earth[26].

In addition to the stretched days in the Old Testament, there are at least two cases in the New Testament where time is distorted in the other direction. In these cases time appears to pass impossibly quickly. First, in John 6:21, Jesus and His disciples immediately arrive at their destination when He climbs into their boat. Second, in Acts 8:39, Philip is transported immediately '*by the Spirit*'.

---

26 I.Velikovski, Worlds in Collision, ISBN 0 575 00230 1, Gollancz.

According to the Bible there will be more disturbances in the future. For example towards the end of the age there will be a huge earthquake, and the heavens and Earth will roll up like a scroll:

> *'...there was a great earthquake; and the sun became black as sackcloth of hair, and the moon became like blood. And the stars of heaven fell to the earth, as a fig tree drops its late figs when it is shaken by a mighty wind. Then the sky receded as a scroll when it is rolled up, and every mountain and island was moved out of its place...'*
>
> **Revelation 6:12-14**

The other future disturbance we will mention is from Zechariah 14:7, which tells us that night will turn into day when Jesus returns and stands on the Mount of Olives. It is interesting that this is opposite to the day turning into night at His crucifixion.

Designed not so long ago

Radiometric dating methods are not reliable

In this section we look at dating methods that are used to estimate evolutionary timescales, and explain how they give us wrong answers with objects and organisms that are older than about five thousand years. Different dating methods are used according to the timescales that are being considered, but all are based on radio-active decay and the technique is called 'radiometric dating'.

The most well-known radiometric dating method uses carbon 14 dating, so let us have a look at the mechanism. The '14' means that there are fourteen particles in the nucleus of the carbon atoms that are used for this method. It is possible to vary the number of neutrons in the nucleus, which does not change the material to a different type, but produces what are called different 'isotopes' of the same material. In the case of carbon, the normal and most abundant isotope is Carbon 12, which has six protons and six neutrons. Carbon 14 is a less common isotope and has six protons and eight neutrons.

This isotope is not stable and in the course of time it 'decays' into a different material, nitrogen gas, and escapes. The time that the decay takes to happen is characterised by a statistical process; by this we mean the time taken for the decay is not constant, but it does have an average value that we know.

The complete process is as follows. First, some atoms of nitrogen are turned into Carbon 14 by collisions with high energy neutrons that arrive from the sun. This liberates one proton and one electron as a side effect from each interaction. The Carbon 14 atoms are absorbed into an organism as part of its normal growing process along with normal carbon (Carbon 12). The Carbon 14 decays back to nitrogen again a few thousand years later. Figure 27 illustrates the process.

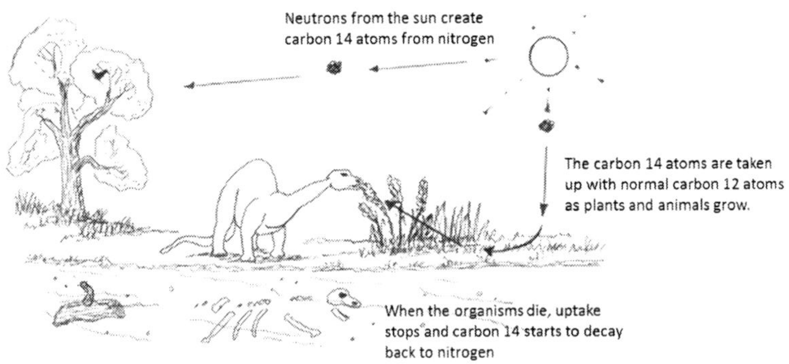

Figure 27. Mechanism by which Carbon 14 isotope is formed and gets built into animals and vegetation as they grow, then starts to decay after the organism dies

In figure 27, incoming neutrons from the sun collide with a few nitrogen atoms and turn them into Carbon 14 atoms.

The radio-active Carbon 14 atoms are absorbed into the tree and the dinosaur as they grow, along with the much more common Carbon 12. When the organism dies, the amount of Carbon 14 present in the organism does not increase any further but begins to decay back to nitrogen. By measuring the ratio of Carbon 14 to Carbon 12 within the dead organism, compared with the assumed ratio when it was living, we can estimate the elapsed time since the organism died. Note that only living organisms can be dated in this way. If something inorganic like a lump of stone or rock needs dating, this can only be done by inference. This means that inorganic dating is performed by dating an organism that is assumed to have existed at the same time.

Alternatively, attempts at dating rocks directly are made by measuring the amount of lead that is present in them. The lead is reckoned to have decayed from the unstable isotopes of Uranium 235 and 238 in a similar way to Carbon 14 decay, but this timescale is over billions of years rather than thousands.

This type of decay is called ‘radio-active’ because, as these materials decay spontaneously, they create some radio interference. Geiger-counters work by detecting this interference. We mentioned earlier that the time taken for this decay is not constant; it is a statistical probability that the decay will happen in a certain time, and this is worthy of a greater explanation for those who perhaps are a little intimidated by the mention of probability.

To illustrate this decay process, imagine that we start with 100 Carbon 14 atoms in a newly dead organism. We observe that the decay has reached 50 percent, or one half, after a certain amount of time has passed. What this means is, after a certain time of say six thousand years, we have 50 remaining of the original 100. After another six thousand years, we have 25 remaining, and after a further six thousand years we have 12 or 13 remaining. Since the number of Carbon 14 atoms halves for each period of six thousand years, we say that the ‘half-life’ of the Carbon 14 isotope is six thousand years. Therefore, since the amount of Carbon 14 in an organism starts to decay into nitrogen when the organism dies, an estimate of how long ago the organism lived can be made by measuring the amount of Carbon 14 that remains not decayed. The exact half-life of Carbon 14 decay is 5730 years.

Have a look at figure 28, this is a graphical representation of the Carbon 14 decay against time as described above. We start with our 100 atoms and after one time period it has dropped to 50, after two time periods it has dropped to 25 and so on. In this graph, the time periods are 5730 years each. The shape of this graph is known as an 'exponential' curve, which is a term you may have come across; all radio-active decay is exponential.

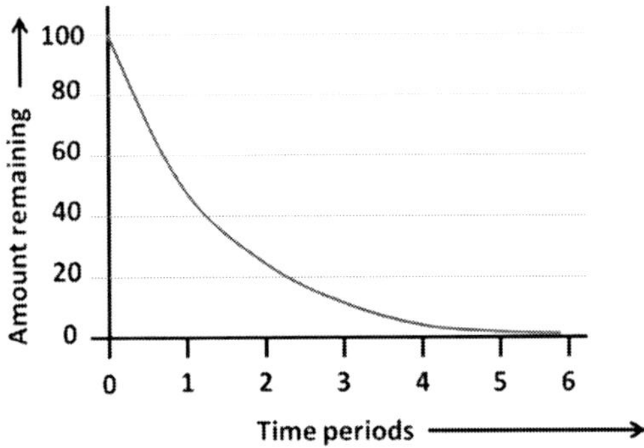

Figure 28. Radio-active decay graph

The time period over which the decay is halved is the 'half-life' of the material. You can see from figure 28 that after four half-lives the line is almost horizontal. The starting time (time = 0 on the graph) is when the organism died, which is what we wish to estimate. If this time were to be estimated from a measurement taken after four half-life periods has passed, the estimate would carry an enormous uncertainty. But despite this, if we assume that four is the maximum number of elapsed time-periods when we can still get a reasonable estimate, then Carbon 14 with its half-life of about six thousand years would only be suitable for dating objects back to a maximum of 24 thousand years. It is certainly useless for measuring millions of years.

Civilisation is known to date back five thousand years and the oldest known objects on the Earth date back to that period, so we can calibrate the any dating method back that far only. For longer periods of interest to the theory of evolution, over millions of years, we have said that Uranium 235 and 238 isotopes are reckoned to be more suitable. They decay into metallic lead at two different rates: the half-lives of Uranium 235 and 238 are estimated at 700 million years and 4.5 billion years respectively and they decay into slightly different types of lead. As these different types of lead can be detected and measured, they provide a cross-check for one another when both types of lead are found in the same rock. Of course it is not possible to calibrate the Uranium 235 or 238 half-lives with just five thousand years of civilisation history, so they are theoretical figures with no support from measurements. It is also not possible to claim that all lead found in rocks was formed by the decay of uranium. Nevertheless some scientists have attempted to estimate the age of rocks on the Earth, and meteorites that arrive from space, from the amount of lead that they contain. Assuming that there was zero lead when the rocks were formed, and the lead they now contain was decayed from uranium, these estimates are typically several billion years. This is similar to the age estimated from secular astronomy as derived from red-shifted light from distant stars, so it has been considered as further confirmation of the 'accepted' age of the universe of billions of years.

For any radio-active dating method to be sufficiently accurate, there are two challenges that need to be met. First, we need to know how much of the radio-active isotope was present in the sample before the decay began. Second, we need to assume that the environment has remained constant where the sample has been kept or has been lying. Neither of these challenges are satisfied because, as we have seen in the previous chapter, there have been cataclysmic changes to the Earth's surface and to the atmosphere. For example we know that, before the world-wide flood, there were no rainbows. If the atmosphere was more protective than it is now, which is highly likely to be the case, it may be that neutrons arriving from the sun were blocked more than they are now.

This means that we cannot know the initial conditions when organisms died, nor can we assume a smooth exponential decay of radio-active isotopes. We conclude that the estimates made by the radiometric dating method can be in error by many orders of magnitude and they should not be believed. What happens in practice is that a very large spread of results is obtained from radiometric dating, and measurement results or observations that do not fit the required theory of evolution are ignored or discarded. This is bad science; the correct procedure is for the theory to be formed around the observations and not the other way around.

Coal and oil

Coal has been demonstrated to form quickly in the laboratory from bio-mass if heat is included in the process[27]. Conditions on the Earth would have been suitable during the flood, where water from underneath and from the canopy would have been heated.

There is far too much of both coal and oil on the Earth for the present land mass to have produced it. In the case of anthracite, which is a very common type of coal, a ratio of 10:1 bio-mass to coal compression is needed. This destroys the swamp theory we were taught at school, because there simply is not room for all the necessary swamps on the current land area. The land would have needed to cover much larger areas than we currently have, to have produced it all. Also many oil and coal deposits are under the sea, for example under the arctic, where there is no bio-mass that could have formed them.

Coal is always found in seams with layers of mud interspersed between the seams. Tidal mud flows can be clearly seen on most of them. A much more plausible explanation than the swamp theory is that coal was formed in the flood, with layers of hot mud compressing the bio-mass.

---

[27] Coal: evidence for a young Earth. See http://www.tgm.org/Creation_coal.html

Significant amounts of Carbon 14 are found in all of our coal and oil and indeed in all carbon-based deposits on the Earth, which means they cannot be millions of years old as some would claim. We saw earlier that Carbon 14 decays to almost zero after twenty thousand years. This leads us to the thought that our coal and oil were formed during or soon after the flood, less than five thousand years ago. It is likely that there was more land area before the flood, which would have supplied the required bio-mass.

The sun and the moon

Our sun is a medium-size star that is burning hydrogen at the rate of about 600 million tonnes per second and converting most of it into helium. It loses about 20 million tonnes in mass every second in the process; we know this because we can measure its energy output.

At the rate that the sun is burning hydrogen, it is shrinking today by about five feet per hour. So it would have been double its present size even only one million years ago. Also, the rate of shrinkage would have been faster in the past due to the need to keep the forces in the sun balanced. This means that even a few million years ago, it would have been larger than the solar system, so it could not have been around for over four billion years as some scientists would have us believe. We have heard one idea that the sun is burning hydrogen faster now and hence is hotter than in the past, but there is no evidence or scientific justification for this.

Regarding the moon, it is gradually moving away from the Earth because it is gradually speeding up. The mechanism for this is as follows. The moon produces a small bulge in the surface of the Earth, which then moves ahead of the moon due to the rotation of the Earth. This has the effect of slowly accelerating the moon and, to keep energy conserved, the moon is moving further away. So the moon is constantly stealing a small amount of energy from the Earth. Let us look briefly at the implications of this. If we go back six thousand years, the moon would have been about 250 metres closer to the Earth (this is not much of a change, considering the moon is 384,400km away on average). The theory that the Earth and moon are over four billion years old, as secular astronomers claim, has a big problem: the moon would have been so close that it would have been touching the Earth less than 1.5 billion years ago.

The Earth and moon form a revolving system that is gyroscopic and hence 'stiff', resisting disturbances. If an asteroid passes close to the Earth, it has less effect because of the presence of the moon. The moon also produces the tides on the Earth, it affects our weather systems that keep the Earth clean, and it irrigates our ocean food systems. We therefore suggest that the Earth-moon system was designed to benefit us and did not come about by chance. It is highly probable that we would not be here if the moon did not exist.

The planets

The planets are very diverse. Let's have a look at them, starting with the planet that is nearest to the sun, and working outwards. Mercury turns on its axis at the same rate as it orbits the sun, so it always presents the same face to the sun. Therefore this face is always in daylight. It has no atmosphere and no moons. Venus has retrograde spin (this means it spins the opposite way to most of the rest of the planets) and a very dense atmosphere of carbon dioxide, nitrogen and water vapour. It has no moons. The temperature of the Venusian surface is several hundred degrees centigrade and scientists think that the dense atmosphere traps the heat from the sun.

The Earth has an atmosphere composed mainly of nitrogen, oxygen and argon, is very hot inside, and has one moon. Mars has a very thin atmosphere of carbon dioxide, nitrogen and argon and has two tiny moons. Jupiter, Saturn, Uranus and Neptune all have atmospheres of hydrogen, helium and methane. They are all gas giants. Jupiter has four main moons, Saturn has one big moon (Titan) and many rings, and Uranus has retrograde spin and six moons. Neptune has one big moon (Triton), and has such a high degree of tilt that it is almost rolling upon its side. Pluto, which has recently been re-classified as a dwarf planet, has a moon almost the same size as itself (Charon).

The planets have different atmospheres, different sizes, different numbers and sizes of moons, differing degrees of tilt, and spin different ways. Some are gas and some are rock. At school we were taught how the planets were formed by coalescing dust and debris that began to spin around the sun, called the theory of accretion. In turn, the moons around the planets are supposed to have formed from coalescing dust that was orbiting the planets in the early solar system. We are told that the evolution of the planets was finished when the solar wind blew all the remaining dust away.

The accretion theory has several problems. If it were true we would expect all the planets to have the same direction of spin and to be composed of similar materials, but they are not. Even if we accept that accretion can produce planets composed of different materials, we would expect any gas ones to be innermost and the denser rocky ones to be outermost, but they are just about the opposite. Another problem is that the angular momentum of the sun is far too low; the sun should be spinning much faster than it is for the planets to have been formed from a spinning cloud around it.

Astronomers tell us that young stars have been discovered in the universe with cool clouds of material surrounding them, and that this supports the theory of planets coming about by accretion. Astronomers are also reporting results of detecting planets around other stars, but the only ones discovered so far are giant gas ones that are relatively close to their star, which is to be expected if they formed through accretion. However we do not believe that accretion can form an array of planets like those in our solar system. The sun is the only known star that has an array of planets including rocky ones.

We suggest that the solar system is an example of design. Mathematical modelling shows that the inner rocky planets are stabilised by the outer gas ones, in a similar fashion to the action of moons that stabilise the planets due to gyroscopic action[28]. Here on Earth we are well protected; we have the sun on the inside of us and a bunch of giant planets on the outside of us. The planets keep us stable, protected, and just the right distance from the sun. Jupiter has been observed to absorb potentially harmful objects that stray into our solar system, attracting them away from the Earth. In 1994, Jupiter attracted and absorbed the comet Shumaker. The planet's gravity attracted the comet into itself and this saved us from potential danger.

The Bible does not say much about the planets, only that they are there as a legacy that is '*allotted to all the peoples under the whole heaven*' in Deuteronomy 4:19.

As well as displaying evidence that they are designed as a system, the planets also testify to the universe being young. Many of the planets of the solar system have strong magnetic fields or magnetospheres, created by electrical currents within their cores that decay over time. The outer planets of the solar system have especially strong magnetic fields. We can measure the decay of the Earth's magnetic field and that of other planets, and we observe that they get weaker every year. If the planets were several billion years old, then their magnetospheres should be extremely weak by now. To account for the present readings and yet say billions of years have passed would mean that their magnetospheres were once impossibly strong. We concede that an exception is Mars, which used to have a magnetosphere but this has decayed to zero. Even taking Mars into account, we assert that the planets are only a few thousand years old, just as the Bible teaches.

Disorder increases with time

The next problem for evolution that we describe is that disorder increases with time. We have already come across this in relation to the flatness of the universe in chapter 7 when dealing with the big bang theory. Disorder is the opposite of order. Anything that is moving or changing in any way will tend to get more disordered,

---

[28] See for example http://en.wikipedia.org/wiki/Planetary_habitability

or more chaotic, with time. For example, if I put a carrier-bag that contains potatoes in the boot of my car without tying up the top, and drive the car around, within a short period of time many will have come out of the bag and arrived in different corners of the boot. The situation at the beginning, with all potatoes in the bag, has a certain amount of disorder, but at the end of my journey the amount of disorder has increased. Another example is my daughter's bedroom; it tends to get untidier as time passes. If she tidies up her room then the order increases locally, but some of her effort is expended in accomplishing this. She consumes energy, and when we consider the calories she has used, and the wider system of food production to provide those calories, the overall disorder in the universe still increases. So, we can reduce the disorder at certain confined locations where we have jurisdiction, such as a room, but the overall disorder in the universe is continually increasing.

It is well-known in science that disorder is increasing in the universe, obeying a law called the second law of thermodynamics. It is related to the fact that the universe is expanding (or being spread out) as we have already discussed, because when anything expands it is becoming more disordered, unpredictable. This law is also demonstrated by gases whose motion becomes more randomised as they spread out, for example when sprayed from an air freshener aerosol can. So in the universe disorder is increasing, which means that the whole of creation is becoming more and more disordered.

There is a relationship between disorder and loss of information. Strictly, the degree of coupling between disorder and information depends upon the statistical nature of the disorder, which we will not go into here. As order in the universe overall is decreasing, chaos is increasing and information is being lost. It is not possible to gain information in a system where the disorder is increasing and, as disorder is increasing in the universe, it is not possible to gain information. This has devastating implications for evolution as we will see.

A believer in evolution might argue that order and information can increase in a local area of the universe, such as the Earth, while across the overall universe they are being lost. This is a bit like tidying up a room but on a larger scale, which requires organisation across the whole Earth just as it did across the room. Such organisation is beyond the capability of mankind. Note that decreasing order and loss of information is actually consistent with the way creation works. It means that the human race started off as good but has been deteriorating and, without divine intervention, will end up in Hell. This is exactly what has been happening since Adam sinned as described in chapter 6. Mankind's deterioration that leads to death is illustrated in figure 29, where he is going along the wide path down towards the grave, led by his decisions and actions. Jesus offers a different path.

Figure 29. The sin of Adam put mankind on a path of increasing disorder to Hell, a wide path that goes downhill. Jesus gives us the opportunity to get off this path onto an ordered one, a narrow path, a divine escape from the second law of thermodynamics

Mankind does not need to go down the pathway of his own decisions and actions towards the grave, but this is the one he will naturally follow in alignment with the scientific second law of thermodynamics. Jesus referred to this as a wide path that most people take. Jesus gives us the opportunity to get off this natural path and onto a divine path where order and information increases. This path is shown in figure 29 as a narrow exit going uphill, the one Jesus said that few would find. Jesus injects order into those who believe, probably without causing an increase in disorder in the rest of the universe. The order He supplies would come directly from Heaven.

However the second law of thermodynamics is not consistent with the theory of evolution, which says that species become higher or more ordered over time. To enable a higher species to evolve from another one, an increase in information would be needed for the increased functionality. Yet we know that if many copies are made of anything, information gets lost rather than gained. Anyone who has made photocopies of a document with many photocopy generations will know this from experience. The outcome gets more fuzzy, with more errors and less order, as copies of copies are made.

So the theory of evolution assumes that information and order increases with time, which is not in agreement with science. Have a look at the situation in figure 30.

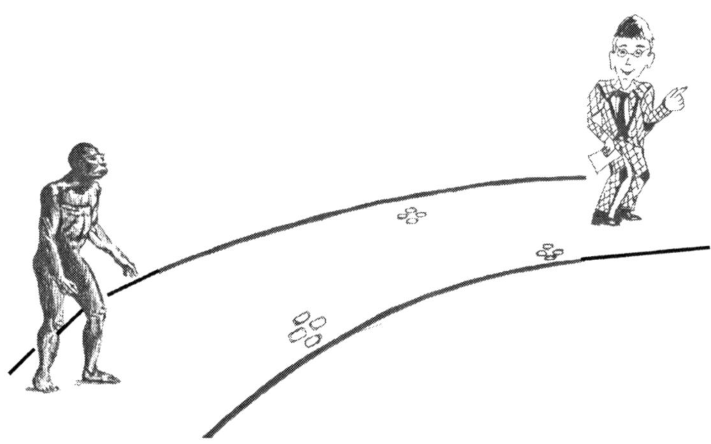

Figure 30. Evolution says that mankind has progressed and developed in a good way with order increasing with time, but this is contradictory to science

Additional information is needed for new species to develop from existing ones, especially if the new species is more advanced. There is nowhere that this necessary additional information can come from. We assert that the passing of time can result in loss of information but not in increase of information, and therefore evolution of the type shown in figure 30 has not happened.

Another theory that we will briefly discuss is theistic evolution. This theory says that God used evolution as a tool to develop human life, so that evolution was a planned process where lower species already contained the information necessary to form higher ones. The theory is that God put into some species all the ingredients and the processes to allow them to evolve into higher species over a long time. We dismiss this theory because we do not believe it is scientifically feasible for primitive life-forms to hold all the information that would be necessary to build higher life forms. It is also not in accordance with scripture. Below are three verses from Genesis that tell us each species reproduces only after its kind; respectively the plants, the sea creatures, and the land creatures:

> *'Then God said, 'Let the earth bring forth grass, the herb that yields seed, and the fruit tree that yields fruit according to its kind, whose seed is in itself, on the earth'; and it was so...'*
>
> **Genesis 1:11**

> *'Then God said, 'Let the waters abound with an abundance of living creatures, and let birds fly above the earth across the face of the firmament of the heavens.' So God created great sea creatures and every living thing that moves, with which the waters abounded, according to their kind, and every winged bird according to its kind. And God saw that it was good...'*
>
> **Genesis 1:20-21**

> *'Then God said, 'Let the earth bring forth the living creature according to its kind: cattle and creeping thing*

and beast of the earth, each according to its kind'; and it was so...'

GENESIS 1:24

The scriptures above tell us that all species reproduce after their kind and therefore cannot produce species of any other kind. Also, Genesis 1:1 to 2:3 clearly states that the whole creation was completed in six literal days. All the plants and animals, and finally humans, were created during that time, so no new species have appeared since. On the seventh day, God rested because He had finished.

Mutations and variation, micro and macro evolution

In this section we draw important distinctions between mutations and variation, and between micro- and macro-evolution.

When an organism reproduces, mistakes are occasionally made that lead to mutations, all of which cause defects. Common mutations in the human population include Down's syndrome, cyclopia, and mermaid syndrome. Although upsetting for families, the sad truth is that such mutations usually result in a short life and a significantly reduced probability of succeeding in finding a mate, and hence of reproducing.

Such mutations are caused by faulty genes. Genes are the parts of DNA within cells, and they carry the codes (or blueprint plans) for manufacture of proteins that in turn are used for building cells in the body. Faulty genes can be directly inherited from either parent, or can come about through mistakes in copying when cells divide as the organism grows. We believe that all mutations we observe are detrimental to the individuals concerned, they are never beneficial. In contrast, variations in individuals are normal and do occur, which sometimes gives an individual a benefit over the others. An example is a particular colour of fur or feathers being a more effective camouflage.

There are reports on evolution websites of a few mutations that have been observed to produce new genes that are of benefit to individuals, and of mutations in metabolic pathways to enable, for example, plants to do better in different lighting conditions. These are cited by some people as evidence of evolution and there is always lots of debate on the Internet. However, these mutations do not result in a different species being formed, but in a perhaps more radical variation in the same species. We would argue that it is variations and not mutations that are being observed. We challenge anyone to prove us wrong by showing us solid evidence of just one beneficial mutation in an animal.

Let us look a little closer at how variations occur. When parents get together to produce offspring, each parent contributes half of each of several DNA molecules, called gametes, which become the chromosomes of the new individual, a chromosome being a part of the DNA thread. The gametes are produced in the parents by a process called meiosis, where a surprisingly complex process produces eggs or sperm with a large amount of genetic variation. In the case of humans, each parent contributes twenty-three gametes to form twenty-three pairs of chromosomes in one huge DNA chain, and a complete copy of this is contained inside every cell in the body. So every cell in the body contains the 'blueprint' for the entire body, but only certain parts of the blueprint are activated in each different type of cell. Stem cells are types of cell that can differentiate into other types of cell, and there are various types of stem cells, classified in terms of their ability to differentiate. At the top of the list of stem cells is a fertilised egg; it is a single stem cell that can differentiate to form any of the cells needed for the embryo to develop.

When cells divide, the new cells contain a complete DNA copy and sometimes mistakes are made in the copying, leading to information loss and, in some cases, to mutations. We repeat the fundamental point that information can be lost in this copying process but it cannot be gained. Each chromosome carries several thousand genes, and different genes are active in different types of cell, for example brain cells have a different set of active genes than stomach cells. It is slight but normal variations in these genes that cause variations in characteristics such as size, colour of skin or fur, and size and colour of eyes.

Technical details aside, the point we are making is that growth and maintenance of an organism depends upon accurate copying of DNA which, if it goes wrong, will usually result in some loss of fitness in the individual affected. This reduces the probability of attracting a mate, and hence any offspring being produced. It is a mechanism that is designed to discourage propagation of faulty genes. Likewise, any variation that is of benefit in the environment will increase the fitness of the individual. As we know from our dating experiences, the fittest individuals are more successful in the competition to attract a mate and hence pass on any increased fitness to the offspring. This is part of the natural selection process.

Such variation and natural selection results in offspring that are more adapted to their environment, but no new information is made available during the process. From these thoughts we identify a distinction between 'micro-evolution' and 'macro-evolution', as follows.

Micro-evolution is where a species changes some characteristics in the longer term, caused by variations in normal genes, through natural selection and inheritance. This causes changes in particular characteristics such as colouring, or loss of some function that has become unnecessary due to changes in environment or lifestyle. We do observe this happening. Reversions are also observed. Macro-evolution is where a new species develops from another one (or other ones). This is not observed.

Now, an evolutionist will try to mix the two. Charles Darwin observed that some variations can be beneficial. An individual possessing a beneficial variation has an advantage over others in the population. He or she will attract a fertile mate and pass on the advantage to the offspring through inheritance. In particular, if the environment is changing slowly, such as getting hotter, or the food supply is changing in character, certain variations will tend to give some individuals an advantage over others for survival; they will be fitter. Individuals that happen to be born with advantageous variations will breed more successfully than those without them. Such variations might be in camouflage so they don't get eaten so readily in a new environment, or in shape of teeth or claws to help them deal better with a new type of food. We are not saying that a change in environment or food causes the variations, but that the variations will happen anyway and some will give advantage to the individuals depending on how the environment is changing. This is natural selection of the fittest and in this way populations can adapt to or 'track' their environment. So far so good, we do observe this happening; this mechanism increases the probability of the species surviving. Examples include our flightless birds, and the pepper moth that adapts its colour to match those of the trees in its environment.

Darwin also observed that variation among domesticated breeds is much greater than variation among wild breeds, for both animals and plants. Our domestic dogs, for instance, show much more variety than wild dogs. Domestic birds show much greater variety than wild ones. Budgies come in a great variety of colours, hardly any two are the same, but the blackbird shows very little variation between individuals, and the natural selection mechanism keeps the characteristics of the wild birds very tightly controlled. Any slight variation usually puts an individual wild bird at a disadvantage, so that the variation does not get passed on to any offspring. Darwin discovered a cause for the difference in variation between domestic and wild species, which is that the abundance of food affects the sexual reproduction organs. He found that if there is plenty of food available, the sexual organs produce gametes that result in a greater variety of offspring than if food is scarce. This is a very valuable finding by Darwin and he did a good job in analysing variation and its relationship to the environment, including the food supply.

While he was a naturalist on a ship called the Beagle on a voyage to the Galapagos Islands, he observed two populations of birds that had started out as one population but got split into two, with the two parts living on different islands that had different environments. The two populations had adapted by displaying different variations. The mistake Darwin made, possibly on the advice of his friend Lyell, was to extrapolate this situation by proposing that, if enough time passes, the bird populations on the two islands would become sufficiently different that they would become different species. This is the crux of his argument in the book 'On the Origin of Species'.

We assert that such an extrapolation is wrong; no matter how much time passes or how much the populations adapt differently to different environments, they are still the same species. Our assertion is based on the lack of new information. The two populations may be considered sub-species, similar to the notion of human 'race', but they are still the same species.

Just to be clear on what a species is, the definition of two organisms being the same species is that they produce fertile offspring. The fact that your parents produced you is proof that *their* parents were of the same species. It doesn't matter if they were from different races, for example one Chinese and one African. An example of non-fertile offspring from different species is the mule, which comes from a horse and a donkey. Mules are infertile; every time you want one you have to mate a horse with a donkey. Hybrid plants are similar.

This notion of Darwin's, of new species arising from variations in two populations over enough time in different environments, was seized upon by people who were very keen to do away with the need for a designer or creator. The work of Darwin is still used as a basis for the argument that new species could evolve from other species. We repeat that no matter how much time passes, it is impossible for one species to evolve from another one, because there is no new information forthcoming in evolutionary processes. There is only information loss and the occasional mutation. You may like to argue that mutations can be beneficial as mentioned before, but there is no evidence of even one mutation that has resulted in a new species, let alone the many hundreds that would have been needed for mankind to evolve from primitive life forms.

Fossils

We showed in the last chapter that fossils were formed quickly and not very long ago, certainly within the last ten thousand years. In this section we revisit fossils to show that they support the account of creation, and provide further evidence against the theory of evolution.

We have been taught at school and we constantly get told by nature programs that we evolved from primates in a progression that took four million years, something like that shown in figure 31 (cell phone excepted).

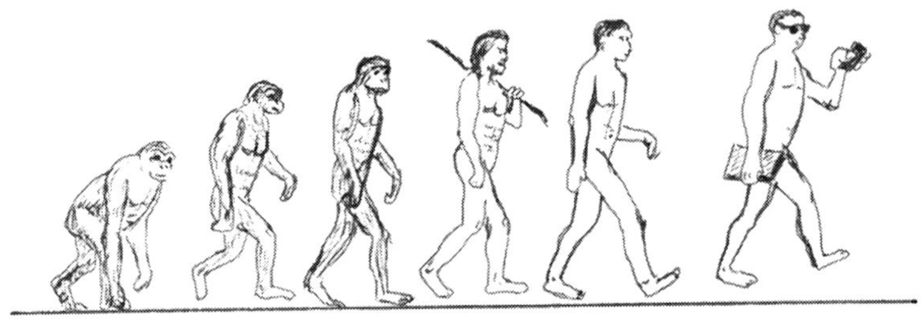

Figure 31. We did not come about like this

A believer in evolution may argue that there is evidence for a common ancestor of apes and humans. If this was really the case, considering all the possible branching of species, and of these only a small percentage would have been successful, we should have an abundance of fossils of transitional forms. However there are very few fossils of transitional forms and these are highly dubious. Among them there is none that can be proven to come from a common ancestor.

A skeleton from an ape was unearthed in 2002, and researchers have recently assigned a new genus and species to it (Pierolapithecus catalaunicus), stating that it 'could be the last common ancestor of modern great apes: chimpanzees, orang-utans, bonobos, gorillas, and humans'. [29]

[29] News bureau of the University of Missouri, May 2013

The reasoning behind this argument is that the shape of its pelvis indicates that it was from an early ape age, before the apes began to diversify. This conclusion depends upon very many unprovable assumptions, such as there was diversification in the first place and that such diversification included a path to evolution of humans. This fossil has not been dated, but the age is estimated at 18 million years based only upon the pelvis and skull shape. This estimate of age is not reliable, as it is based upon previous assumptions about the evolutionary timescale, which we have seen are flawed themselves.

We argue that there is no credible fossil or other evidence of such a common ancestor or of any resulting diversification. There just is not sufficient fossil evidence of common ancestors or transitional forms to support the theory.

The Cambrian Explosion is a significant event in palaeontology that is well documented. This 'explosion of fossils' divides prehistoric times into pre-Cambrian and Cambrian and a sketch is shown in figure 32.

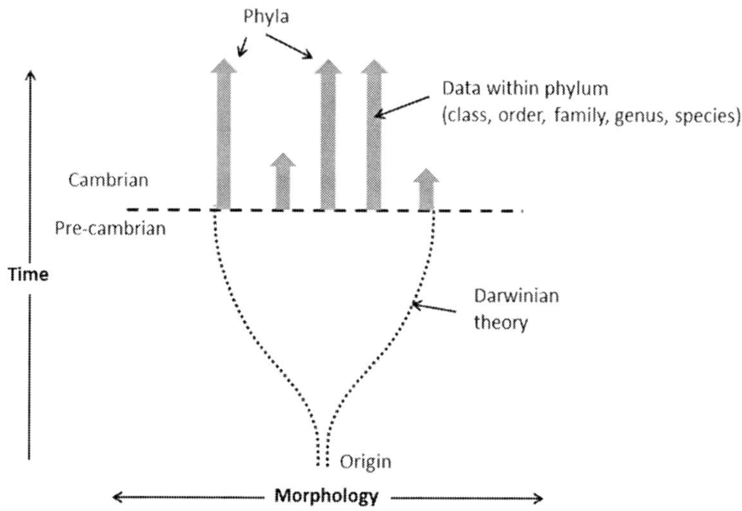

Figure 32. Sketch of the Cambrian Explosion, usually dated at 550 million years ago

The Cambrian Explosion represents a time when many millions of fossils were formed, which represent all known species of plants and animals that are around today and also some unknown and extinct ones. The usual timescale put on this explosion is 550 million years ago. Before this time, in the pre-Cambrian, species are supposed to have become distinct from each other through evolutionary processes over a largely unknown time, back to the beginning of life on Earth.

As indicated in figure 32, the only fossil or other evidence that we have available is from after the explosion; there is none available from before. Towards the bottom of the figure, in the so-called pre-Cambrian period, the theory becomes increasingly murky and unsubstantiated, relying upon the Darwinian theory of diversification of species as described earlier.

The origin of life, how it all started, depicted at the extreme bottom of the figure, is a mystery to evolution scientists. Much money and effort is being expended on trying to establish a plausible story. The current candidate stories include life originating through hydrocarbons arriving on a meteorite from outer space, life coming about from lightning flashes into a kind of primordial soup, and life starting from carbon compounds that form near thermal vents in the oceans.

The Cambrian Explosion is itself an embarrassment to the evolution theory. There is an abundance of evidence for this explosion happening, with most of the earliest fossils dating back to this time. The fossils from this time show no evidence of any species changing at all, let alone gradually evolving into higher species.

We deduce that at the Cambrian Explosion all the species started out together, and have not changed since, just as the book of Genesis said they did. So we agree about the explosion happening, but would disagree about the time it happened; we believe it happened a few thousand years ago rather than 550 million years ago. We have already seen how such long timescales are estimated through the use of radiometric dating, and how they cannot be trusted.

Flies and butterflies

Some insects have multiple stages in their reproductive cycles. Some plants such as the fern also have two or more stages in their reproductive cycles. These organisms are not a good fit into the evolutionary process of survival of the fittest. An example is the fly; its reproductive cycle starts with the egg, then larvae, then three stages of pupae and finally the adult. Survival is important at all stages. Any variation that gives an individual an advantage at one stage will not necessarily benefit the individual at a later stage.

Another example is the butterfly with its intermediate caterpillar stage a shown in figure 33.

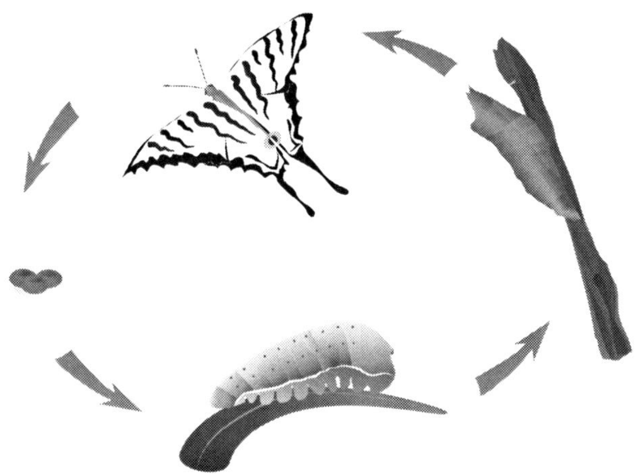

Figure 33. Life cycle of butterfly[30]

Any characteristics possessed by the caterpillar that make it more likely to survive do not necessarily help the second stage where it is important to find food and a mate, and find them quickly. Likewise, any variation in the second stage that helps an adult butterfly to survive and quickly find food and a mate does

[30] Image credit Designua/Shutterstock

not necessarily increase the fitness of the caterpillars that come about as a result of their union.

Number of offspring and no modification over time

We have briefly investigated the number of offspring that different organisms produce. In table 2 we have put flies and butterflies already mentioned and two larger organisms, tigers and humans.

Organism	Offspring development time	Number of offspring in lifetime of parents
Fruit flies	7 days	About 2000
Butterflies	A few months	About 700
Tigers	100 days	About 30
Humans	280 days	Usually less than 8

Table 2. Offspring development time and number of offspring for some example organisms

You can see from table 2 that the more complex organisms typically take a longer time to produce offspring and that they also have fewer offspring. If evolution were correct, then the higher organisms would need more variations and more mutations, requiring more and more sophistication, to get where they are today. To support this, the number of offspring would need to increase, but we observe the opposite: higher organisms have fewer offspring.

We also observe that, despite the very much greater numbers of offspring produced by insects per generation, which would increase the likelihood of mutations, they have not changed at all since the earliest observations. Ants caught in amber five thousand years ago have exactly the same form, function and social behaviour as today's ants.

In fact, no species that is still living has been observed to change at all, not from the time of the earliest fossils, from the time of the so-called Cambrian Explosion.

This evidence tells us two things. First, no new species has ever been observed to develop or evolve from another species and, second, no species that is alive today has changed noticeably from the earliest fossil records of that species. This evidence witnesses against the theory of evolution and instead supports exactly what is written in the creation account in the Bible:

> *'God made all sorts of wild animals, livestock, and small animals, each able to produce offspring of the same kind. And God saw that it was good'.*
>
> GENESIS 1:12

Dependent systems

We use the word 'dependent' to describe a system where some parts depend on other parts, and only when the system is complete does it become beneficial to the organism. It is not possible that these systems evolved gradually, because any part of the system would be a handicap to the organism. The complete system has to be ready all at once. Let's look at some examples.

Cells

Living bodies are made up from many different kinds of cells. A cell has many parts or components, which need to have been ready all at the same time in order that the cell can work. A basic picture of a cell is shown in figure 34.

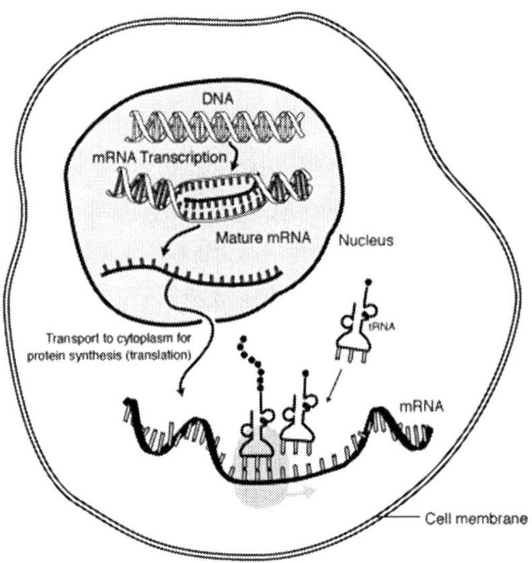

Figure 34. The cell has many parts that depend on each other, so it must have been designed as a system

We have come across DNA and genes already from the earlier discussion on variation. Genes are the parts of the DNA that are responsible for building the cells and differentiating them from other cells. For example a hair cell is different from a skin cell and will have a specific colour and texture, and the genes are responsible for making a cell a hair cell, and include the coding for its colour and texture.

Figure 34 shows the process of protein building in the cell, enabling the cell to grow and fulfil its function. Building the proteins is a two-step process. The first step is called transcription, where messenger ribonucleic acid (m-RNA) forms along the gene. It matures in a complicated way that we won't describe here and then is transported to the ribosome. The second step is called translation where the m-RNA fragment is used to code the construction of a protein, within the ribosome. The m-RNA is called messenger because it carries information from the DNA to code the protein.

The proteins are constructed from amino acids that are supplied from the digestion of food, transported in the bloodstream and eventually brought to the ribosomes within the cell by tRNA (t means transport).

The point we are making is that protein building in the cell is a complicated process that involves several steps and components, where some of the steps depend upon others. It is not possible that any of these steps, or components, on their own, could benefit the individual, in fact the opposite would be true; any component or subset of the system on its own would be a handicap, and natural selection would get rid of it. You need the combination, all components working together as a system, to be of benefit. So we assert that the cell could not have evolved; it must have been designed as a system.

Blood clotting

Another example of a designed system is the process of blood clotting. This process, with its seven steps, is shown in figure 35. We shall not describe the process, but notice from the figure that certain steps require the presence of a chemical, much like a catalyst, to succeed. You can see from the figure that two pathways converge into one, depending on the source of the damage that initiates the forming of the clot. The probability of this system emerging through random variations, while not putting individuals at a disadvantage with a partially developed system, is beyond credibility. So we conclude that the process could not have evolved gradually; it must have been designed.

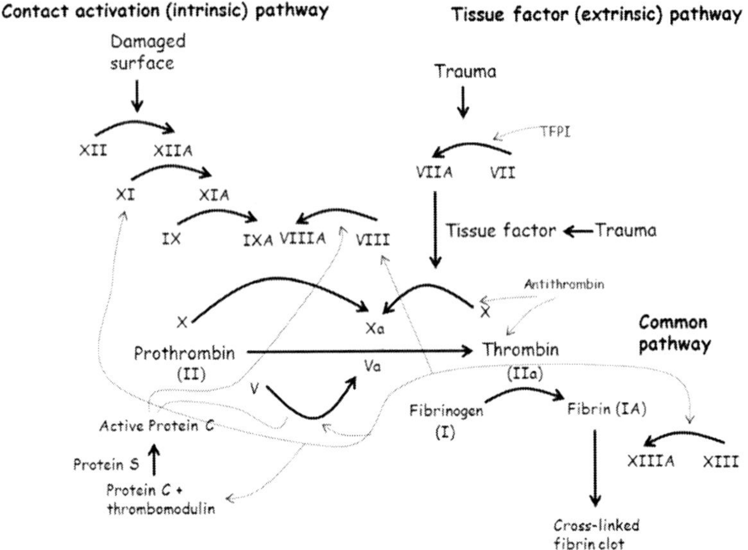

Figure 35. The blood clotting process is another system that surely must have been designed

Flagellum

The human sperm and many species of bacteria use a flagellum, a kind of revolving tail, to propel themselves along. The bacteria that have these are supposed to pre-date man; they are found in some of the earliest fossil records. Figure 36 shows the component parts of a flagellum.

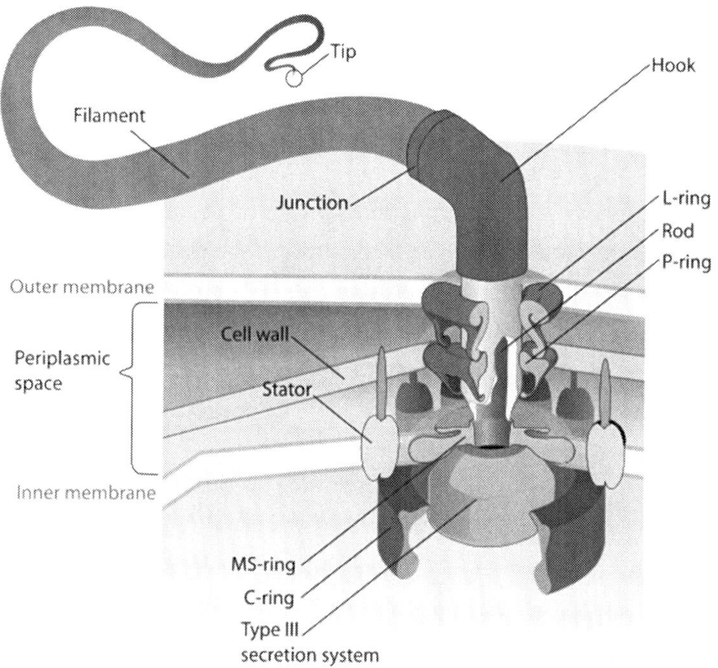

Figure 36. The flagellum such as that on the human sperm is made of fourteen components that work as a system

The flagellum is made from different proteins and resembles the design of an electric motor. It even has a clutch arrangement. All fourteen components are required for it to operate; any subset or partial system would be a handicap to an individual, worse than useless. So this motor could not have gradually evolved; it must have been designed.

Sex and reproduction

The mention of sperm above leads naturally to a discussion about sex. Sexual reproduction requires that both male and female systems were ready together. Maintenance of the sexual apparatus in human males and females is very costly in terms of energy and body processes, much more so than asexual apparatus, because of the continuous production of eggs and sperm, most of which are wasted.

This system cannot have evolved gradually, as it would have been an enormous handicap for an asexual organism to develop any part of the sexual apparatus in isolation. Such a partial development would have put the organism at a disadvantage. The organism would not have won and succeeded in an environment where food and mates were competitive, hence it would not have passed on the partial development to offspring to build upon. We deduce that sex did not evolve, but was designed so that the male and female parts were ready as systems and also together. After all, you can't reproduce before you can reproduce! Recall the discussion earlier about Adam being created male and female, and then the female parts were removed and formed into a woman.

The eye

The final system that we consider is the eye. The structure of the human eye is shown in figure 37.

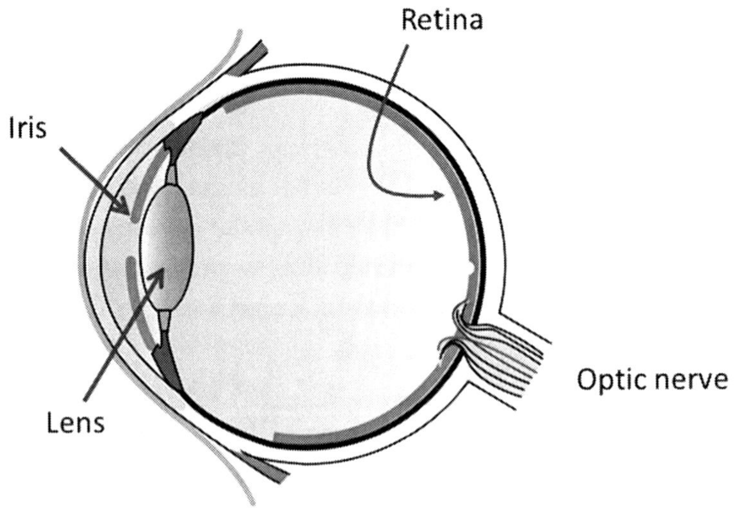

Figure 37. The human eye[31]

[31] Image credit Dragana Gerasimoski/Shutterstock

The positioning of the photo-sensitive retina behind a highly complex adjustable lens and iris system is a marvel in engineering. Darwin argued in his book 'On the Origin of Species' that all you need is a piece of tissue that happens to be sensitive to light for an eye to appear by the process of genetic variation. We have heard the theory that over hundreds of thousands of generations, a sheet of light sensitive cells gradually curved up and formed the eyeball. The theory goes on to propose that, in front of the light sensitive cells, a chance lump of gunk gave an individual the ability to focus an image. This combination is said to have evolved into the variable lens we have now.

This gradual evolution of the eye is said to be proven from examples of eyes of marine animals, from the most basic light sensitive cells present on worms, right up to the most complex on the octopus that has eyes much like ours. Whereas there are marine animals with different types of eyes, we assert that they were all designed and they are not examples of the eye in different stages of evolution. A lump of gunk might have attached itself by chance to a light sensitive cell on one individual, but it would have come from outside and not appear as a random variation or mutation. Such a structural change would require new information, which is not possible to acquire through copying. But even if we imagine for a moment that a lens could evolve like this, it is beyond belief that this was accompanied by further random mutations that led to arrangements of pairs of eyes being in exactly the right place on heads. A pair of eyes is needed to judge distance accurately and also to compensate for the 'blind spot' that each eye has at the place where the optic nerve leaves the retina. Surely this came about by design.

Some may argue that the eye is a highly evolved piece of equipment, but in this case one would expect to find them on the later fossils only. But even the earliest fossils have eyes; some have very complex ones with multiple compound lenses, implying that they were designed like this when they were created.

Why some will not give up on macro evolution

We have made the case that there is no evidence for macro-evolution, or one species evolving into another one. There is no direct observation of it happening and there is no evidence in the fossil records. Indeed we observe the very opposite, species are becoming extinct before our very eyes. Yet still the western governments, national broadcasters and teaching establishments hang on to the macro-evolution theory. We were taught it at school. Our children are still being taught it at school. On-going research activity all over the world is trying to force it to make sense. Why is this? Why the desperation to hang onto a wrong theory? It is very unscientific to maintain a theory when there is a lack of evidence to support it, and much evidence that it cannot explain.

We said in the introduction that good scientific practice is to propose and maintain a theory based on observations. If observations are made that do not fit the theory, the theory must be modified or replaced with a new one. An example of this was discussed in chapter 7 on the universe, where the size of an object is described by classical three-dimensional theory. This is accurate enough for most applications, but for very small or very fast objects, we observe effects that cannot be explained by the classical theory. New theories were proposed, quantum theory and relativity, to explain the new observations. Quantum theory and relativity are two of the most successful scientific theories, but even these will need to be modified or replaced if observations are made that do not accord with them.

We said that this process does not seem to have happened to the theory of evolution. We have now seen that there are many counter observations, such as the Cambrian Explosion, fossil evidence for giants upon the Earth, coal and oil with Carbon 14 present, and the dependent systems. We said in the introduction that evolution was promoted from a scientific theory to a scientific law, which seems to make it immune to counter observations. The American Scientist magazine describes evolution as an 'established scientific law', and so it is very well established in the minds of scientists on the payrolls of universities and governments. Millions of dollars of funding is poured into evolutionary research every year. We can only think that these researchers are interested in maintaining their research grants, but moreover they want to find a solution that avoids the question of creation (and hence of a Creator). Governments do not want the Bible to have any credibility as a source of information about how or why life came about.

Evolution over millions of years, with humans as the highest of the evolved beings, is a comfortable thought. A very slowly changing environment on the Earth and everywhere else in the universe is another comfortable thought. We have challenged this by showing that the environment on the Earth has not been stable; it has been subject to many sudden changes such as meteorite impacts, orbital disturbances, and a flood. We encourage you to look for yourself for further evidence of these cataclysms; it is to be found everywhere. We have also shown that there exists a great deal of evidence that the universe is less than ten thousand years old, and that everything came about through design and not through evolution.

Restoration

Where we are in the timeline

In the fullness of time (as time appears to us), God will restore all things to Himself. Restoration is a process that consists of many stages, and some of them have already taken place. It will go beyond what we are able to imagine in terms of time and relationship with God. The creation will return to what God originally intended it to be; the new Heaven and the new Earth will be without the presence of sin or the sin nature. People who are saved will spend on-going ages with God, even beyond what we are told about in Revelation. Paul's letter to the Ephesians tells us that God will point to believers as examples of His grace in coming ages:

> *'And God raised us up with Christ and seated us with him in the heavenly realms in Christ Jesus, in order that in the coming ages he might show the incomparable riches of his grace, expressed in his kindness to us in Christ Jesus.'*
>
> EPHESIANS 2:6-7

The first step in the restoration process, which we have witnessed as already taken place, is that Jesus was the first to be born from the dead, and the first to receive eternal life in a resurrected body. He was dead and separated from God when He became the curse, as we have seen.

We are told that Jesus overcame the world before He went to the cross. Believers in Christ are given the same Spirit that raised Jesus from the dead in order that, like Him, we can overcome the sin nature within ourselves. The reward for overcoming is that we will be raised with Him and inherit eternal life. It is virtually impossible to deal with the sin nature alone; it is highly beneficial to join a community of believers because it reduces the chance of going astray. One of the principal roles of the church, which is such a community of believers, is to deal with the sin nature. The three scriptures below should encourage you to believe:

> *'But if the Spirit of Him who raised Jesus from the dead dwells in you, He who raised Christ from the dead will also give life to your mortal bodies through His Spirit who dwells in you.'*
>
> **Romans 8:11**

> *'These things I have spoken to you, that in Me you may have peace. In the world you will have tribulation; but be of good cheer, I have overcome the world.'*
>
> **John 16:33**

> *'Most assuredly, I say to you, he who hears My word and believes in Him who sent Me has everlasting life, and shall not come into judgment, but has passed from death into life.'*
>
> **John 5:24**

Jesus has been given spiritual authority and secular authority, but for now He has entrusted the secular authority to His body the church:

> *'Go therefore and make disciples of all the nations, baptizing them in the name of the Father and of the Son and of the Holy Spirit'*
>
> **Matthew 28:19**

He remains the invisible head of the church, and is acting through it as a kind of proxy. When the church has fulfilled the great commission in the above scripture, it will be taken up in the rapture. At this time He will begin assert His secular authority directly and, when He does, the pace of restoration will increase.

Let us take a look at Daniel chapter 2, which describes a dream that king Nebuchadnezzar had. It will help to clarify where we are in relation to the sequence of empires. This dream was spiritual and prophetic, and it disturbed Nebuchadnezzar so much that he searched among his magicians to try and find one who could interpret it. None of them could do so. In a series of miracles, God revealed the dream itself and its meaning to the prophet Daniel, who proceeded to reveal it to the king. The dream was of a giant statue. The head of the statue was gold, representing Babylon and Nebuchadnezzar personally. The chest and arms were of silver, representing the Medes and Persian empires that were to follow. The belly and thighs were bronze representing the Greek empire. The legs were of iron representing the Roman empire. The feet were of iron and clay mixed, representing democracy and bureaucracy that followed. The feet had ten toes, the number ten meaning decisions or democracy. This last empire is the new Babylon, a new world order in the last days. Notice that the make-up of the statue from head to feet goes from rare to plentiful, expensive to cheap, soft and malleable to hard and brittle. It has an interesting parallel with the world in which we live, where things are becoming increasingly plentiful, cheap, and brittle. In the dream, the statue is destroyed by a huge rough stone that strikes its feet, and then this stone gets larger and larger until it takes over the whole earth. The rough stone points to Jesus and His kingdom that will be set up.

The name Nebuchadnezzar means 'O Nebo defend my defend my empire and crown'. Nebuchadnezzar became king of Babylon in the year 604BC after the death of his father Nabopolassar. Nebuchadnezzar captured Jerusalem in 597BC; he attacked king Jehoaikim and carried him off to Babylon along with some of the treasures from the temple that Solomon built. Nebuchadnezzar installed Zedekiah as king, but in 589BC Zedekiah unwisely formed an alliance with Egypt. Nebuchadnezzar responded by invading Judah and laying siege to Jerusalem that lasted about thirty months. Zedekiah's eyes were gouged out and he was carried off to Babylon along with the remainder of the temple treasures, then Jerusalem was razed to the ground and the temple destroyed.

In Revelation chapter 17 a woman and a beast are described, where the beast has seven heads and ten horns. The seven heads represent seven mountains (empires), and the ten horns represent ten kings. The seven empires are described in Revelation 17:7-14 as Caldese, Egypt, Babylon, Medes and Persians, Greece, Romans, and the Antichrist. This is in agreement with the kingdoms represented in Nebuchadnezzar's dream, with the Antichrist mapping onto the feet with their ten toes made of clay and iron, indicating that the Antichrist will take power from the bureaucratic law. It the dream, the large stone strikes at the feet, a prophecy of the Christ defeating the kingdom set up by the Antichrist.

So the seven empires in Revelation are the fulfilment of the prophetic dream of Nebuchadnezzar; presently here on Earth we have come through the first six, and number seven is being set up. These seven have and will influence the whole world, seven meaning completion. Patterns from the first six have influenced kingdoms today even to the technology that the Antichrist will use in number seven. Kingdom number seven is very near to being finally established.

When kingdom seven is established, we will be in a period known as the tribulation, which is described in more detail in the next section. A new world order will be in place whereby, in order to buy or sell anything, everyone must carry what is called in Revelation 13:17 '*the mark of the beast*'. The beast could be the name of the global trading system that will be in place, its computer system, or perhaps even the nick-name of its leader, the Antichrist.

A mark on the forehead identifies the system that the person belongs to, and also identifies the lifestyle and character involved, like a kind of branding. There are different types of these marks, which clearly identify the allegiance of people. The people of God have such marks according to Deuteronomy 6:8, and the 144,000 have the seal of the Father on their foreheads according to Revelation 7:2.

With the mark of the beast we pick up the trail of the global trading theme that we left behind in chapter 3, where widespread trading is stated as the reason for Lucifer's downfall. Trading is a hallmark and a repeat of the cause of Satan's original downfall, and it will also lead to his final downfall in the end. Such trading is a characteristic of Babylon. The global trading system will possibly be derived from today's banks or computer servers belonging to banks. This system could well be the harlot who rides the seven-headed beast in Revelation 17.

The Antichrist is also known as the man of perdition and the man of lawlessness. He may also be called the beast because he is following Satan and devours like a beast for his own purposes. There is some debate about whether the beast is given this secular power before spiritual power.

The global trading system or new world order will be deployed to control everyone, and people who refuse to submit to it will be liable to punishment or death. The mark could be a bar-code etched into the forehead or on the hand or a chip placed under the skin:

> *'He causes all, both small and great, rich and poor, free and slave, to receive a mark on their right hand or on their foreheads, and that no one may buy or sell except one who has the mark or the name of the beast, or the number of his name.'*
>
> **Revelation 13:16-17**

The technology to do this is available right now. Maybe first it will be introduced in an optional way to make transactions easier, but then it will become a requirement to have it to trade at all. Please stoutly resist taking such a mark or device into on onto your body. The risk of starving to death or being put into prison is preferable to the consequences of accepting the mark. We have heard people say that we may not realise when we have accepted the mark but, as we have said, we believe the mark will be one of allegiance to be taken deliberately, so there is no danger of taking it without realising what it is.

If you are in doubt about the consequences of accepting the mark, consider these scriptures from Revelation:

> *'If anyone worships the beast and his image, and receives his mark on his forehead or on his hand, he himself shall also drink of the wine of the wrath of God, which is poured out full strength into the cup of His indignation. He shall be tormented with fire and brimstone in the presence of the holy angels and in the presence of the Lamb'*
>
> **Revelation 14:9-10**

and

> *'...the smoke of their torment ascends forever and ever; and they have no rest day or night, who worship the beast and his image, and whoever receives the mark of his name.'*
>
> **Revelation 14:11**

We repeat, please don't ever take this mark; it will lead to your destruction at the final judgement. It would be better to starve to death than to take the mark.

Returning to the subject of where we are in the timeline right now, you can see around us that kingdom seven is being established. Trading in shares and commodities is global in nature. Trading houses such as stock exchanges are striving for ever shortening delays in trading transactions, even to the point of shaving off microseconds. This is so that money can be made if one trading centre is lagging slightly behind another one when shares are rising or falling. Dealers buy from one and sell to the other. Trading centres across the United States and across Europe are connected together using microwave radio links, rather than fibre optic links, because delays of microwave links are a few microseconds shorter.

It is dangerous to try to ascertain the exact timescale for when the kingdom of the Antichrist will be fully established. Many have tried to calculate it, and there have been some embarrassing predictions of the end of the world by different religious cults. Nevertheless an idea can be gleaned from these words from Jesus:

> *'Assuredly, I say to you, this generation will by no means pass away till all these things take place…'*
>
> **MATTHEW 24:34**

Here Jesus is talking about events that occur during the tribulation. By '*this generation*', we assume He means people who were alive when Israel was established in 1948 by UN Resolution. As most people do not live longer than about ninety years we would be surprised if it hadn't happened by 2038. But we are told that the exact time is known only by the Father as explained in Matthew:

> *'But of that day and hour knows no man, no, not the angels of heaven, but my Father only'*
>
> **MATTHEW 24:36**

and

> *'In such an hour as you think not the Son of man comes'*
>
> MATTHEW 24:44

The rapture and the tribulation

In the book of Thessalonians, we are told this:

> *'For the mystery of lawlessness is already at work; only He who now restrains will do so until He is taken out of the way. And then the lawless one will be revealed, whom the Lord will consume with the breath of His mouth and destroy with the brightness of His coming. The coming of the lawless one is according to the working of Satan, with all power, signs, and lying wonders...'*
>
> 2 THESSALONIANS 2:7 – 9

The one who restrains is the Holy Spirit within believers, who are described as salt and light; these are commonly used for preservative and illumination, according to Matthew 5:13-16. We said in chapter 2 that the Holy Spirit is deposited within people when they become believers, so that believers in Christ are temples where the Holy Spirit currently dwells. The Holy Spirit will be taken out of the way when believers are suddenly removed from the Earth in the rapture.

Only those who are prepared and walking in the Spirit will be taken. The story of the five wise and five foolish virgins in Matthew 25 tells us that the five foolish virgins did not have enough oil in their lamps to keep them burning when the bridegroom was delayed and arrived later than expected. The wise virgins had extra oil with them and were admitted into the wedding banquet when the bridegroom finally arrived, while at that moment the foolish ones were away getting more oil. When the foolish ones returned, the wise ones had been taken inside and the door was shut. The message here is, be prepared by being filled with more of the Holy Spirit than you think you need.

We have said that those who walk with Jesus are the temple of the Holy Spirit; so when believers are taken in the rapture this temple will be gone. The Antichrist will allow a replacement temple to be established on the Temple Mount in Jerusalem, possibly the one being planned by the Temple Institute we mentioned earlier. The Dome of the Rock and Al-Aqsa mosques and other Islamic buildings are currently in this location; we can only think that these will have to make way for the new temple. It will be interesting to see how this is achieved.

The removal or taking of believers at the time of the rapture will be a life-changing event for everyone, and will make headline news in every nation. Believers will suddenly just disappear:

> *'Then shall two be in the field; the one shall be taken, and the other left. Two women shall be grinding at the mill; the one shall be taken, and the other left. Watch therefore: for ye know not what hour your Lord doth come.'*
>
> **Matthew 24:40-41**

Believers that have died and believers who are still living will be taken up to meet the Lord in the air:

> *'For this we say to you by the Word of the Lord, that we who are alive and remain until the coming of the Lord will by no means precede those who are asleep. For the Lord Himself will descend from heaven with a shout, with the voice of an archangel, and with the trumpet of God. And the dead in Christ will rise first. Then we who are alive and remain shall be caught up*

> *together with them in the clouds to meet the Lord in the air. And thus we shall always be with the Lord.'*
>
> 1 THESSALONIANS 4:15-17

Believers who are still alive will be changed in a '*twinkling of an eye*' at the time of the rapture. We will be equipped with heavenly bodies suited to their new environment:

> *'we shall all be changed – in a moment, in the twinkling of an eye, at the last trumpet. For the trumpet will sound, and the dead will be raised incorruptible, and we shall be changed.'*
>
> 1 CORINTHIANS 15:50 – 52

This scripture refers to a '*last trumpet*', which is the last call to the church; this is different to the last trumpet in Revelation 11:15. The one in Revelation heralds the start of God pouring out His wrath on the Earth in the great tribulation.

Although the rapture or taking of the church from the Earth may seem far-fetched, especially to anyone who is not a believer, it is not the first or the last time that a rapture (or ascension into Heaven) has occurred or will occur again. There are three incidents recorded in the Bible where people have already been taken miraculously into Heaven, and two more will happen in the future. The three already raptured are Enoch, Elijah, and Jesus. The two to come are the rapture of the Church that we are discussing here, and the future rapture of the Two Witnesses who we will be introducing shortly.

The rapture is not to be confused with the second coming of Jesus; they are not the same thing. In the rapture Jesus comes for His church, but does not set foot upon the Earth; the church meets Him in the air. In the second coming, He returns for the Jews and stays to reign on Earth for one thousand years (the millennial reign).

After the rapture most people who remain on the Earth, those who were not part of the church, will be initially very happy. There will no longer be any church trying to evangelise or putting any restraint on sin and immorality. People will be free to do as they wish without any conviction of guilt. It is highly likely that Christmas and Easter celebrations will be abandoned. Then again some people will be devastated because they thought they were believers but were not taken. These people will probably set up communities to work out their salvation another way, which will be possible although with a high cost, as we shall see later.

There is some debate about the timing of the rapture; some are convinced that it will occur at the beginning of the seven year period called the tribulation, others say it will occur in the middle, and yet others say it will occur at the end. We confess that we are of the first group; we believe that the church will be taken before the tribulation begins. The scriptures relating to the rapture do not immediately appear to agree with each other, which has led to some confusion. The confusion is remedied if we conduct an investigation into who the various prophecies are talking about. One example of where confusion can occur is where Jesus says this about the great tribulation:

> *'And unless those days were shortened, no flesh would be saved; but for the elect's sake those days will be shortened.'*
>
> **Matthew 24:22**

It is tempting to interpret the *'elect'* here as the church, indeed some Bible versions translate this as 'chosen ones', but this creates a difficulty if we interpret this verse as talking of the tribulation after the church has been raptured. The difficulty vanishes if we interpret the elect in this scripture as the 144,000 Israelites that are sealed in Revelation 7.

Another example where confusion can happen is where Jesus says this:

> *'Therefore when you see the 'abomination of desolation... standing in the holy place' - then let those who are in Judea flee to the mountains. Let him who is on the housetop not go down to take anything out of his house. And let him who is in the field not go back to get his clothes. But woe to those who are pregnant and to those who are nursing babies in those days! And pray that your flight may not be in winter or on the Sabbath.'*
>
> **MATTHEW 24:15-20**

We can wonder who Jesus is talking about here. It cannot be the church that should run for their lives when they see the abomination of desolation standing in the holy place, if the church has already gone. This difficulty disappears if we consider that Jesus is talking about the Jews, as is implied by Him telling them to pray that their flight is not on the Sabbath.

Paul emphasises the absence of the church from the whole tribulation period by saying:

> *'For God did not appoint us [the church] to wrath, but to obtain salvation through our Lord Jesus Christ'*
>
> **1 THESSALONIANS 5:7**

The wrath spoken of here is God's judgment on the unbelieving world and His discipline of Israel during the tribulation. One purpose of the tribulation is to bring the Jews back to God.

The 'abomination of desolation' is probably where the Antichrist sets up the deliberate sacrificing of pigs in the new temple in Jerusalem in the future. In 167BC, a Greek ruler called Antiochus Epiphanies set up an altar to Zeus in Jerusalem and sacrificed a pig, and this event became known as the 'abomination of desolation'. When Jesus later used the same term in Matthew 24:15, we assume He meant that there will be a repeat of this act.

So the scriptures lead us to believe that the rapture will be at the beginning of the seven year period, and the church will not endure any part of God's wrath. Indeed it seems that the taking of the church actually triggers the start of the tribulation. As all believers will be taken before the tribulation, the implication is that there will be no warning of the rapture. There will be no signs to tell us that it is about to happen, apart from the signs we currently see, such as the world banking and trading systems descending into increasing dis-array waiting for a leader to be revealed. Paul confirms that there will be no warning:

> *'But concerning the times and the seasons, brethren, you have no need that I should write to you. For you yourselves know perfectly that the day of the Lord so comes as a thief in the night. For when they say, 'Peace and safety!' then sudden destruction comes upon them, as labour pains upon a pregnant woman. And they shall not escape. But you, brethren, are not in darkness, so that this Day should overtake you as a thief...'*
>
> **1 Thessalonians 5:1-4**

Destruction will come unexpectedly upon those who are not ready, just like it came upon the five foolish virgins. Jesus will suddenly call His believers up like a thief in the night, and time is getting very short. Believers had better be ready from now on.

The covenant of grace will end at the rapture, meaning that it will be very much more difficult afterwards to obtain salvation and eternal life. It will still be possible, but with great suffering and possibly even requiring the shedding of blood and loss of life. Note that Jesus's mandate to save people will not have changed, but the process will be different. In 2 Thessalonians 2:11 we are told that God will '*send a strong delusion*' after the rapture to make it difficult for people to see the truth, so that only the determined will be saved from Hell by Jesus.

The work of the Holy Spirit will adopt a new working method during the tribulation. Rather than helping people to overcome the world and lead them to Jesus from the inside as He does now, He will work from the outside to encourage those who turn to God. The Holy Spirit goes back to the Old Testament method of being 'on people' rather than 'within people', because the covenant of grace has finished.

Encouragingly though, there are groups of people who make it through the tribulation. One such group is the 144,000 elect from the first fruits of the Jews that we have already mentioned. They have God's seal on their foreheads. Another group is described in Revelation as a 'great multitude' of people from every tribe and nation, who made it through the tribulation:

> *'After these things I looked, and behold, a great multitude which no one could number, of all nations, tribes, peoples, and tongues, standing before the throne and before the Lamb, clothed with white robes, with palm branches in their hands, and crying out with a loud voice, saying, 'Salvation belongs to our God who sits on the throne, and to the Lamb!'*
>
> *These are the ones who come out of the great tribulation, and washed their robes and made them white in the blood of the Lamb.'*
>
> **Revelation 7:9 and 14**

Jesus is often referred to as the sacrificed Lamb of God. The use of the word '*tribes*' in the above scripture reminds us that many Jews who are not the first fruits will be included, and will achieve salvation in this manner.

After the church is taken away in the rapture, the Antichrist will set himself up on the Earth as a great leader, a charismatic person who appears to be interested in world peace and stability. This will be a deception. It is part of the great delusion sent by

God in 2 Thessalonians 2:11 upon those who refuse to trust in Christ. The Antichrist achieves leadership through flattery and high intelligence, beauty and charisma that he has acquired as gifts from Satan. It is apparent that Satan will offer this man the same thing that he offered to Jesus when He was being tempted: *'All these things (kingdoms) I will give You if You will fall down and worship me'* from Matthew 4:9. Unlike Jesus, this man immediately accepts the offer.

After the rapture, the Antichrist will proceed to broker a deal, a peace treaty with Israel. He appears as a prince of peace and the treaty will guarantee peace for seven years. Part of this will be the allowing of the new temple to be built on the Temple Mount in Jerusalem as we said earlier. Note that this temple is not in God's plan for restoration particularly, but it may serve to unite the Jews together on a project they will passionately believe in.

Although the Antichrist is revealed as a person after the Holy Spirit is taken away, he will not reveal his true colours immediately; he will wait for three and half years, half way through the seven year period. He will start his work straight away, but it will not become obvious that this man is actually the Antichrist until the half-way point. He will wait for something else to occur before he makes his move, which we will come to shortly. This delay in the Antichrist asserting himself can be interpreted in 2 Thessalonians 2:6 from the words '*in his own time*'. At the half-way point, he will break the peace treaty with Israel and the second half of the tribulation will start. This is the period known as the 'great tribulation', which we describe in the next sections.

The seven seals and the first six trumpets

In Heaven there will be a courtroom scene. John writing in Revelation says this:

> *'And I saw in the right hand of him that sat on the throne a scroll written within and on the back, sealed with seven seals.'*
>
> **Revelation 5:1**

Here Jesus is increasing His assertion beyond that established in Matthew 28 to now include secular authority, and claiming His inheritance by opening the scroll. The scroll is the title deeds of the Earth that have been recovered from Satan. Recall that Satan took them from Adam at the time of the fall in order to steal the dominion originally given to mankind. The title deeds are now restored to their rightful owner. Jesus is the only one qualified to remove the seals, because He is the Creator, the perfect sacrifice; a lamb without blemish that has been slain.

The events in the courtroom scene in Heaven are triggering the events upon the Earth. During the first half of the seven years tribulation, the first six seals will be removed from the scroll. Removal of the first four seals are the cues for the riding out of the four horsemen of the apocalypse that will trigger wars, famine, pestilence and physical changes to the Earth respectively, as described in Revelation 6:1-8. These are fore-shadows of the ultimate war that the Antichrist will bring in later.

At the removal of the fifth seal, the souls of martyrs will appear under the altar in Heaven. This group is comprised of souls that have given their lives for the sake of the Word of God or have been killed because they refused to deny Jesus and refused to take the mark of the beast. They ask how long it will be before God judges the wicked and avenges their blood. According to Revelation 6:9, the response is that they are issued with white robes, and asked to wait a little while longer until the number of their fellow servants and brothers who were to be killed as they had been is completed. This implies that there is a certain number of martyrs that is required.

When the sixth seal is removed, Revelation 6 describes the scene on Earth:

> *'I looked when He opened the sixth seal, and behold, there was a great earthquake; and the sun became black as sackcloth of hair, and the moon became like blood. And the stars of heaven fell to the earth, as a fig tree drops its late figs when it is shaken by a mighty wind.*

> *Then the sky receded as a scroll when it is rolled up, and every mountain and island was moved out of its place.'*
>
> **REVELATION 6:12-14**

These plagues will be sent against the Egyptian gods, such as the stars falling and the sea that was used for trading turning to blood.

Revelation 7:1-8 tells us that at this time there will be 144,000 Israelites sealed; 12,000 from each of the twelve tribes. These are landowners, which they have been since the days of the twelve sons of Jacob. In fact there were thirteen tribes in the Old Testament, because Jacob adopted Joseph's sons Ephraim and Manasseh and made them both into a tribe. This was Jacob giving Joseph a double portion as a reward for saving the family from famine. So in the Old Testament there were twelve tribes plus a priestly tribe, Levi. This has a parallel with the thirteen people present at the last supper; twelve apostles plus Jesus the priest.

In the list of twelve tribes in Revelation 7, Manasseh is present and Ephraim is called Joseph. The tribe of Dan is missing and the Bible does not say why. One idea is that Dan were cut off from their inheritance because they indulged in idolatry according to Judges 18:30-31.

The list of tribes in Revelation 7 is written out laboriously because it is describing the title deeds of the Earth, and the 144,000 are witnesses to the property ownership. It is almost like the reading of a will. One way of looking at this is that when Jesus died he left the will, the scroll, which is a legal document saying that the land has to be given back to the rightful landowners, who are God and His Christ. The property covered by the title deeds will not be harmed until these witnesses are sealed; the angels that are holding back the four winds are told:

> *'Do not harm the earth, the sea, or the trees till we have sealed the servants of our God on their foreheads.'*
>
> **REVELATION 7:3**

The 144,000 referred to in the above scripture are the elect that live on the Earth during the tribulation, but they are sealed by the Father and keep themselves perfect without sin. They are virgins (literally or spiritually) and they are the first fruits of the Jews. They have the Fathers seal on their foreheads; they are not part of the church nor do they take the mark of the beast. They remain pure witnesses and an offering to God. Revelation says of them:

> *'These are the ones who were not defiled with women, for they are virgins. These are the ones who follow the Lamb wherever He goes. These were redeemed from among men, being first-fruits to God and to the Lamb. And in their mouth was found no deceit, for they are without fault before the throne of God.'*
>
> **Revelation 14: 4-5**

The role of the 144,000 after this point is not explained to us in the Bible; possibly they will help those going through the tribulation as hinted in Daniel 11:33, or they may help to prepare the Jews for the return of Jesus.

The breaking of the seventh seal is followed by a period of silence in Heaven of '*about half an hour*' according to Revelation 8:1, and during this time a golden censer is swung. Then there will start a series of seven trumpets being blown in Heaven. The blowing of the first four trumpets will bring about hail, fire, blood, burning sulphur upon the Earth and also the smiting of the sun. Trumpets five and six will trigger the first two woes or terrors, which are plagues of locusts and horsemen, and more plagues as described in Revelation 9. On Earth, people will witness asteroids and meteorites hitting the Earth, and great changes in the sky such as the sun being smitten. The first six trumpets will be warnings, or heralds, of the gospel and the mercy of God. While they are being blown, there will still be the opportunity to be saved through repentance. The amount of time that passes from the breaking of the first seal, to the end of the sixth trumpet will be three and half years, that is, from the beginning of the seven year treaty with Israel to half way through it.

The seventh trumpet and the great tribulation

The blowing of the seventh trumpet will herald the start of the second three and a half year period that is known as the 'great tribulation'. This is the final trumpet, announcing that God has finally run out of patience; several things happen and we focus upon just three of them.

The first is that a proclamation will be made in Heaven in Revelation 11:15, stating that the kingdoms of the Earth now belong to God and His Christ. They have now been returned to their rightful owners, following the removal of all seven seals. This proclamation will infuriate Satan who chases after a woman who is about to give birth, so he can devour the baby as soon as it is born.

This is the culmination of the battle between the seed of Eve and the seed of Satan. The offspring of the woman, a symbol of Jesus, will be snatched up to Heaven and, realising that the offspring is out of reach, Satan will pursue the woman instead, but then the Earth will open up to protect her. Then war will break out in Heaven where Michael's army will fight Satan's. Satan will fight back but will be defeated and be thrown down to the Earth. Here he will pursue the rest of the woman's offspring, who are those that keep God's command and have the testimony of Jesus (see Revelation 12:17). The people that Satan goes after at this time will be those that are going through the tribulation, who are the sealed 144,000, the Jews, and those who have become believers in Jesus after the rapture.

Recall here that Satan was the lord of the air, the first heaven, but at this point he will become confined to the surface of the Earth as the next stage in him being cast down. When he realises he is confined to the Earth, Satan will have great fury and will pursue those belonging to God and to Christ because he knows his time is short:

> *'Woe to the inhabitants of the earth and the sea! For the devil has come down to you, having great wrath, because he knows that he has a short time.'*
>
> **Revelation 12:12**

So when Satan is thrown to the Earth the evil trinity on Earth will be complete; consisting of Satan, the Antichrist, and the False Prophet, respectively mimicking members of the Holy Trinity; consisting of Father, Son and Holy Spirit. This is what the Antichrist has been waiting for to reveal his true colours.

We have not yet discussed the False Prophet. He is a man who the Antichrist will recruit to handle his public relations, a media-man whose job is to convince the world that the Antichrist is the saviour. Many people will believe that he is the Christ but we are warned about this by Jesus in Matthew 24:24, which says that many false christs and prophets will appear. This will be Satan's last ploy to control the Earth and capture as many souls as possible to join him in his certain destruction. He wants to deceive everyone and the days will be shortened to prevent the elect (144,000) being deceived, as described in Matthew 24:22.

The second event that will be triggered by the seventh trumpet is that God will send the Two Witnesses to walk around Jerusalem for 42 months (three and a half years) according to Revelation 11:3. This means they will be around for the entire second half of the seven year period, which is the great tribulation. Some believe that the Two Witnesses are the resurrection of Elijah and Moses, but we cannot be sure. Unlike the 144,000 who are witnesses to the property, the Two Witnesses represent the dualities of the old and new covenants. These dualities are the Jews and Gentiles, and the law and grace.

During the second half of the tribulation, the Two Witnesses will prophecy the wrath of God. They will be anointed with unprecedented power from God the Father Himself to shut up the sky and to bring plagues upon the Earth as often as they wish, according to Revelation 11:6. They will be very unpopular to the global trading organisation and, after three and a half years, right at the end of the seven year tribulation period, the Two Witnesses will be killed by the Antichrist. Their bodies will be left on display in Jerusalem. There will be a great celebration at their death, and people will give each other presents.

After three and a half days the Two Witnesses will rise from the dead and stand up, which will cause great fear and commotion in Jerusalem. They will be taken up into Heaven (raptured), accompanied by an earthquake that will kill many thousands of people:

> *'Now after the three-and-a-half days the breath of life from God entered them, and they stood on their feet, and great fear fell on those who saw them. And they heard a loud voice from heaven saying to them, 'Come up here.' And they ascended to heaven in a cloud, and their enemies saw them. In the same hour there was a great earthquake, and a tenth of the city fell. In the earthquake seven thousand people were killed, and the rest were afraid…'.*
>
> **REVELATION 11:11-13**

So at the half-way point of the tribulation period, directly after the seventh trumpet is sounded, the evil trinity will be on Earth, and the Two Witnesses will be on Earth with great power. The tribulation moves to the great tribulation and the Antichrist will increase his assertiveness, and he will break the peace treaty with Israel.

He will erect a statue in the temple at Jerusalem, which is the image of the beast. The image will probably be a state-of-the-art piece of technology that will be connected to the trading systems.

It will be given power to speak and utter profanities against Jesus according to Revelation 13:15. The image is given the breath of life by the False Prophet, which mimics the breath of life given to Adam by God. It is another counterfeit.

There is a parallel event here with Nebuchadnezzar, who made a gold image of himself after Daniel told him that he was the king of kings in Daniel chapter 3. Daniel's friends refused to bow down to the image and this got them thrown into a furnace as punishment, but Daniel 3:25-30 tells us that Jesus joined them and saved them from the flames. Similarly we can expect punishments on those who refuse to bow down to the image of the beast and hopefully they will have faith that Jesus will save them from the flames.

The third event triggered by the seventh trumpet is that seven bowls of wrath will be poured out from Heaven, which brings about consequences for the Earth that are described in Revelation 15:1 to 16:21. It would be best to read these scriptures for yourself. God's wrath and judgement will be poured out on the Earth on a scale that cannot be imagined. He really has run out of patience. These are judgements that are punishments for rebelliousness and witchcraft. Towards the end, the human race itself will be in danger of being wiped out.

In stark contrast to what is happening on the Earth, in Heaven there will be preparations made for the marriage of Christ with His bride, with feasting and celebrations. This is a spiritual wedding, where Jesus marries His church, which are the people taken up in the rapture. These people will be all who believe in Christ, both dead and living, from all nations including Jews that have come to know Him as saviour in the period between His ascension and the rapture.

This means that some of mankind will be fully united again with God, which was the condition before the fall of Adam. The wedding itself takes place as described in Revelation 19:7-9. The seven years tribulation is in alignment with the seven days celebration for traditional Jewish weddings in the case where the bride is a virgin. The bridegroom would traditionally arrive at five to midnight, and take her to his father's house to live in a recently built new storey. She is prepared by her attendant virgins or maids. But in this case the bride (the church) has got herself ready, according to Revelation 19:7. During the seven day traditional wedding period, the two families are brought together. Following the marriage will be a period of time when Jesus reigns with His bride the church. The period will be a thousand years, otherwise known as the millennium; almost like a kind of honeymoon.

The second coming of Jesus and the thousand year reign

Towards the end of the great tribulation, on the plains of Megiddo in Israel, nations will be assembled with the objective of wiping out the Jews, and making Satan king over Jerusalem and the world. The evil trinity has orchestrated this because Satan knows that while there is even one Jewish person left living, Jesus will come back for that person. The assembled armies will begin to move towards Jerusalem, which is about 80km to the south. The battle of Armageddon will begin.

Then Jesus will return to the Earth to save the last remnant of the Jews and indeed mankind itself. Like a flash of lightning going from east to west that every eye upon the Earth will see, Jesus will appear and stand on the Mount of Olives across the Kidron valley from Jerusalem. This is known as the second coming.

When He comes, He will proceed to bind up Satan, overthrow the beast's trading system, and throw the Antichrist and the False Prophet alive into the Lake of Fire. This is the *'great and glorious day of the Lord'* as described by Matthew 24:21-22 and Revelation 19:11-21.

The Mount of Olives will split in two from east to west as He walks, as described by Zechariah 14:4-5. A valley will be formed in the Mount of Olives, and people will escape to safety through this valley. Zechariah 14:6 explains that the sources of light will not shine and, although it will be evening, there will be light. On that day there will be an eclipse or some other reason for the sun not shining; possibly it will have been smitten as described in Revelation 8:12. At that time, Jesus himself will be the physical light of the world as well as the spiritual one. When Jesus died on the cross the day was turned into night, and here on His second coming the night will be turned into day.

Appearing with Jesus will be the church that is now His bride, she is His 'completer'. Other covenant groups that will be present and go into the millennium reign with Jesus are those people who will have come to believe in Jesus after the rapture and are still alive, and the sealed 144,000.

From the Mount of Olives, Jesus and this great crowd will walk down to the Jerusalem city wall and through the Golden Gate, a curious mixture of heavenly bodies and earthly bodies. Now at the present time this would be quite difficult, because in between the Mount of Olives and the gate is the Kidron valley and a major road. An additional inconvenience will be that this gate has been bricked up since the days of Ottoman Empire. So either some earthquakes or substantial civil works will be needed to provide a suitable pathway and to unblock the gate.

It is interesting that right now some radical Jews in Jerusalem are planning to construct a viaduct across the Kidron valley at this very location, and also they are planning to rebuild the temple. Indeed all of the furniture has been built to go into a new temple by an organisation called the Temple Institute, including a gold menorah and table of showbread. These Jews are getting ready to storm the temple mount and retrieve it from Islam. They plan to build and equip the temple and construct the viaduct, and begin sacrificing animals again as in Old Testament times.

The appearance on the Earth of Jesus and His saints upon the Mount of Olives will herald the start of the millennium and a process of clearing up and of restoration will begin. This will be a period of peace, stability and joy. Jesus will start the process of separating the sheep from goats on a nationwide basis, using as His criteria those who have accepted or rejected His rule.

According to Ezekiel 39:7-14, it will take seven years to restore the situation and burn the weapons. It will take seven months just to bury the dead. Jesus will proceed to send plagues on the nations that attacked Israel; Zechariah 14:12 describes very graphically how these plagues will cause suffering such as flesh melting, and eyes rotting in their sockets. During the thousand years, all other nations including Egypt will be required to send representatives to Israel and celebrate the Feast of Tabernacles (or Shelters). If they do not comply, no rain will fall on their land as described by Zechariah 14:16-19.

During the millennium, according to Zechariah 8:22-23, the Bible says that non-Jewish nations will seek God, but will have to clutch at the sleeve of a Jew to show them the way. This scripture mentions that ten men of different languages will do this (ten means wisdom); it seems to indicate that a single language will be restored as part of the way, as in earlier times from the creation up to the time of the tower of Babel.

At that time also there will be another modification of the covenant but we do not yet know all of the details. One detail we do know is a change in animal kingdom relating to the food chain. Isaiah 11:4-6 describes an Earth where *'the lion and lamb lay down together and a child will lead them'*. This reminds us that the Lion of the tribe of Judah is now sitting on the throne of the Lamb. Creation begins to return to what was originally intended. Births, marriages and deaths will continue among the human population, and people will be healthy and live long according to Isaiah:

> *'No more shall an infant from there live but a few days, Nor an old man who has not fulfilled his days; For the child shall die one hundred years old...'*
>
> **Isaiah 65:20**

The end of the age

At the end of the thousand year reign Satan will be released for a little while as described by Revelation 20:7. He will occupy his very short respite by deceiving the nations again, and lining them up for one last battle for Jerusalem, where Jesus will be established as king. At the height of this battle, Revelation 20:9 says that fire will come down from Heaven onto the attacking armies and consume them. After this defeat, Satan will be bound again and thrown into the Lake of Fire where the Antichrist and False Prophet already are.

This is followed by the final judgement, referred to in Revelation 20:11 as the 'great white throne judgement' where books will be opened, including the Book of Life. During this process, the souls of people whose names are not found in the Book of Life will be thrown into the Lake of Fire. These are the people who have said 'no' to a relationship with Jesus, and will include all those that have accepted the mark of the beast.

There is a fundamental truth here. Jesus will not come to collect His church in the rapture until the gospel has been preached to every nation. This means that all peoples on Earth will have heard the gospel and been given the chance to receive salvation through Jesus Christ at some point in their lives. Saying no to Jesus means that their names are erased from the Book of Life. At the final judgement they will be thrown into the Lake of Fire for eternity, with Satan, the Antichrist, the False Prophet and all the evil angels for company. This is known as the 'second death'.

At the end of the age there will be a new Heaven and a new Earth as described in Revelation 21 for Jesus and His bride the church, and for the other covenant groups. Upon the new Earth will be the New Jerusalem, a holy city where God Himself will dwell with His people. There will be no more pain or upset of any kind, only untold ages of living as God intended right at the beginning before sin entered; full of love, peace and joy. The tree of knowledge of good and evil will be there as a reminder of the power of sin and the sinful age; people will look at it and be grateful for what God has done. This time there will be no serpent that will come and cause problems. Restoration is complete.

The old Earth will contain Hell for ever and ever[32], a burning prison full of hate, anger, and weeping; it will contain the evil trinity, all the evil angels, the spirits of the Nephilim, any other demons, and the souls of all the humans that Satan has managed to steal.

As for the various covenant groups, there will be the church, the souls who made it through the great tribulation, the 144,000 elect, the Old Testament saints, and maybe others. The future role of the church is to rule and reign with Christ according to 1 Corinthians 6:2-3 with an ever increasing scope of government and peace according to Isaiah 9:7. As for the other groups, we are not certain about their futures or their final roles. We have come to the end of Revelation in the Bible. We may speculate perhaps that the role of the elect in the new Earth is to serve as priests in the temple.

We believe that the new Heaven and new Earth are themselves followed by other yet untold ages, when we are even closer to God through more future covenants and future creations. And of course, within the authority of Jesus, future choices.

---

[32] Literally ages and ages

Epilogue

On the Earth, our way of life will be changed radically, possibly quite soon. Most of us are presently lured into a false sense of security that it won't or can't happen.

Probably most of us enjoy living. The world definitely has pleasures that should be enjoyed, but there is a danger that we become entrapped by them. It is normal to derive pleasure from family, food, power, possessions, sex, and so on, but it is important to avoid becoming pleasure seekers. It leads to wanting more and more, as we have seen with the world lately: obesity, corruption, child abuse, broken families and murder are all on the increase. It will continue to increase.

Especially we all know that there is pleasure in sinful activities, such as adultery, gluttony, revenge, and dishonest gain, but these pleasures are only fleeting according to Hebrews 11:25. For example divorce commonly follows sexual sin, and the personal experience of one of the authors is that there are no blessings for anyone involved in one.

If we continue on a sinful path, we are heading for Hell. How are we to be saved from this? The good news is that there is salvation from Hell through Jesus, as stated many times already in this book. If a person has a personal relationship with Jesus and confesses that He is Lord, then that person is guaranteed salvation and inherits eternal life with Him. In John 16:33, Jesus tells us we should take heart because He has overcome the world with all its pleasures and threats and dangers. A relationship with Him changes us from pleasure seekers into joy seekers. The world has no hold on Him. A promise given in Revelation says this:

> *'He who overcomes, I will grant to him to sit down with me on my throne, as I also overcame and sat down with my Father on His throne.'*
>
> **Revelation 3:21**

Whatever our lifestyle and behaviour, we cannot ever be good enough to make it into Heaven on our own. If we don't make it into Heaven there is only one alternative, which is Hell. The difference between them is that Heaven is full of love and fulfilment and Hell is full of hate and torment.

We cannot emphasise enough the gravity of the situation and unavoidable choices that mankind is facing. Please be convinced that Satan wants people to be relaxed and unaware of how critical is the choice before us and, if this book has raised your awareness of the choice, it has achieved its aim.

The choice is this: you have to lose your sinful natural life one way or another, make no mistake there will be no escape. Everyone will bow the knee to Jesus. There are just three stark options facing us all.

First and best is to choose to accept Jesus into your life right now. Go and speak to someone else you know who is a Christian or seek out a church and speak to one of their leaders. Then you will start to tap into His abundant grace, and you will overcome your sinful nature with the help of the church and the Holy Spirit who will move in to live within you. Immediately you will begin to lead a new, fulfilled, and joyful life in God, and inherit eternal life. It will never be easier to be saved from Hell than it is now under the covenant of grace.

The second choice is to delay a decision until nearer to the rapture but this is a dangerous tactic because we don't know the exact timing. There will be no warning. Jesus says He will come '*like a thief in the night*' when least expected and some will be caught unprepared. If you miss the rapture, you face the prospect of bringing yourself through the tribulation without the help of the Holy Spirit within you. The covenant of grace will be finished. There is a way into eternal life from here if you manage to recognise what is going on, reject the mark of the beast and acquire faith in God. This route involves much more suffering, and then possibly being killed for your faith.

The third choice, which the Bible tragically says that most people will take, is to reject Jesus, get caught up in the new world order, accept the mark of the beast, not recognise what is going on and lose your life in the great white throne judgement. You will be forced to bow to Jesus, and be condemned to an eternity in the Lake of Fire in the old Earth with the evil trinity, evil angels, and demons for company.

This is not about us telling you what to do, but to put these options before you and explain the consequences of your choice. These things are prophesied in the Bible. There are 747 events prophesied in the Bible. Around 82 percent of these have already come to pass, proved by history and archaeology. The remainder are all in the future, and we have every reason to believe these will also come to pass. They include the coming of the Antichrist, the rapture, the second coming of Jesus, Armageddon, and the new Heaven and new Earth that we have been discussing.

We hope and pray earnestly that you take the first choice above. Amen.

Lightning Source UK Ltd.
Milton Keynes UK
UKOW04f0002050214

225869UK00001B/7/P